W0091514

Piet Mondrian "Tree," 1913.
ⓒ2001 Artists Rights Society (ARS), New York/Beeldrecht, Amsterdam

Progress in Mathematics
Volume 176

Series Editors
Hyman Bass
Joseph Oesterlé
Alan Weinstein

Hyman Bass
Alexander Lubotzky

Tree Lattices

with Appendices by
H. Bass, L. Carbone,
A. Lubotzky, G. Rosenberg, and J. Tits

Birkhäuser
Boston • Basel • Berlin

Hyman Bass
Department of Mathematics
University of Michigan
Ann Arbor, MI 48109-1109
U.S.A.

Alexander Lubotzky
Department of Mathematics
Hebrew University
Jerusalem 91904
Israel

Library of Congress Cataloging-in-Publication Data

Bass, Hyman, 1932-
    Tree lattices / Hyman Bass, Alexander Lubotzky ; with appendices by Hyman Bass ...
[et al.].
       p. cm.– (Progress in mathematics ; v. 176)
    Includes bibliographical references and index.
    ISBN 0-8176-4120-3
       1. Trees (Graph theory) 2. Lie groups. I. Lubotzky, Alexander, 1956- II. Title. III.
Progress in mathematics (Boston, Mass.) ; vol. 176.

    QA166.2.B38 2000
    511'.52–dc21
                                   00-059911
                                       CIP

AMS MR Classification Numbers: 20E08, 20E06, 20E18, 05C25, 05C05, 20D05, 20H05, 22E40,
                57M15

Printed on acid-free paper.
© 2001 Birkhäuser Boston

Birkhäuser

All rights reserved. This work may not be translated or copied in whole or in part without the written
permission of the publisher (Birkhäuser Boston, c/o Springer-Verlag New York, Inc., 175 Fifth Avenue,
New York, NY 10010, USA), except for brief excerpts in connection with reviews or scholarly analysis.
Use in connection with any form of information storage and retrieval, electronic adaptation, computer
software, or by similar or dissimilar methodology now known or hereafter developed is forbidden.
The use of general descriptive names, trade names, trademarks, etc., in this publication, even if the former
are not especially identified, is not to be taken as a sign that such names, as understood by the Trade Marks
and Merchandise Marks Act, may accordingly be used freely by anyone.

ISBN 0-8176-4120-3   SPIN 19901609
ISBN 3-7643-4120-3

Typeset by John Spiegelman, Philadelphia, PA and by T_EXniques, Inc., Cambridge, MA.
Printed and bound by Hamilton Printing, Rensselaer, NY.
Printed in the United States of America.

9 8 7 6 5 4 3 2 1

Dedicated to our roots and branches

in memory of my grandmother
טובה בליזובסקי

A.L.

with love to my children,

Anne, Ivan and Gabriella
H.B.

and in admiring tribute to

J.-P. Serre,

Photo by Lisa Carbone, Jerusalem, Israel, May 2000

Contents

Preface

Let H be a non compact simple Lie group. In most cases H can be realized as essentially the full isometry group of some geometry X, where X is a contractible metric space. This geometric action is the natural tool for studying H and its discrete subgroups, in particular its lattices.

In the case of real or complex Lie groups, X is the symmetric space, a Riemannian manifold. In case H is a Lie group over a non archimedian (for example a p-adic) field, X is the Bruhat–Tits building, a contractible simplicial complex of dimension equal to the rank of H.

There is one notable exception to the above picture, namely when H is non archimedean of rank one, in which case X is a bi-regular tree. In this case the full isometry group $G = \mathrm{Aut}(X)$ is vastly larger than its subgroup H. Nonetheless Tits has shown that G itself is a virtually simple locally compact group, and one can naturally study its discrete subgroups and lattices. That is one purpose of the present work.

More generally we study the automorphism group $G = \mathrm{Aut}(X)$ of any locally finite tree X, and its discrete subgroups Γ. We call Γ an X-lattice if the volume of $X \mathrm{mod} \Gamma$, suitably defined, is finite. These are the tree lattices of the book's title. We present a fairly systematic investigation of the existence, structure, and properties of tree lattices, drawing parallels and contrasts, whenever possible with the situation for lattices in Lie groups. We give much attention to the construction of diverse examples. The methods used are based on the notion of graphs of groups, as first developed in the book *Trees* of Serre.

While the theory of lattices in Lie groups motivates many of the questions investigated here, we do not rely on that theory for the present work, which is essentially elementary and self contained.

Many of the results of the book appear here for the first time in print. This work has evolved slowly over about a decade. Some of the main results presented or cited appeared first in early drafts as conjectures. There are three appendices. One describes a group theoretic construction provided by Peter Neumann which we use here for producing certain self normalizing non uniform tree lattices. A second appendix, by one of us with J. Tits, presents a criterion for the full automorphism group of a tree to be discrete. The third, by one of us with L. Carbone and G. Rosenberg, presents the proof that, whenever G is unimodular and $G \backslash X$ has finite volume, there exist X-lattices. Initially this was only conjectured, and proved in special cases. We are grateful to the other authors for permission to publish these results here.

Finally, we owe a great debt to Ann Kostant and Elizabeth Loew for generous and skillful editorial and production assistance in bringing this manuscript to final completion.

Hyman Bass
Alexander Lubotzky
Jerusalem, July, 2000

Tree Lattices

Chapter 0

Introduction

0.1 Tree lattices

Let X be a locally finite tree. Then $G = \mathrm{Aut}(X)$ is a locally compact group. The vertex stabilizers G_x are open and compact (in fact, profinite). A subgroup $\Gamma \leq G$ is *discrete* if Γ_x is finite for some, and hence every vertex $x \in VX$. In this case we define

$$(1) \qquad \mathrm{Vol}(\Gamma \backslash\backslash X) := \sum_{x \in \Gamma \backslash VX} \frac{1}{|\Gamma_x|},$$

call Γ an *X-lattice* if $\mathrm{Vol}(\Gamma \backslash\backslash X) < \infty$, and call Γ a *uniform X-lattice* if $\Gamma \backslash X$ is a finite graph. In case $G \backslash X$ is finite, this is equivalent to Γ being a lattice (resp., uniform lattice) in the locally compact group G. These tree lattices are the object of study in this work. The technique used is the theory of graphs of groups ([S], Ch. I), as elaborated in [B3]. The study of uniform tree lattices was initiated in [BK]. The present work, in some ways a sequel to [BK], focuses much more on the non-uniform case. Here the phenomena are much more complex and varied. Accordingly, we devote a great deal of attention to the construction and analysis of diverse examples.

We are guided to a large extent by comparisons with the following classical source of examples (and applications). Let F be a non-archimedean local field, with residue field $\mathbf{F}_q$, of cardinal q. Let H be a simple F-rank 1 algebraic group over F. Then H has a Bruhat–Tits tree $X \cong X_{q+1,q'+1}$ (biregular of degrees $q + 1$ and $q' + 1$) on which H acts with quotient $H \backslash X = \circ\!\!-\!\!\!-\!\!\circ$. Then a subgroup $\Gamma \leq H$ is a (uniform) H-*lattice* iff Γ is a (uniform) X-lattice.

We have not treated questions about the spectral theory, in particular the zeta functions of non-uniform lattices (cf. [B4]). We hope to take this up at a later time.

The methods developed here have recently been extended to a study of (irreducible) lattices on a product of two (or more) trees, as a first step toward generalizing the theory of lattices in semi-simple non-archimedean Lie groups. (See [M3], [BG],

[BM2], [BM3], [BMZ], and [G]). In these cases, as expected, one meets rigidity phenomena not present for tree lattices, as well as remarkable examples of uniform lattices on a product of two trees which are simple groups!

Acknowledgements

Shahar Mozes provided much helpful feedback and was a constant source of ideas and insight. Jacques Tits kindly consented to publish here, as Appendix [BT], "The Discreteness Criterion for Tree Automorphism Groups." Peter Neumann supplied the group theoretic ingredients, published here as Appendix [PN], that permitted us to construct self normalizing, non-uniform lattices Γ on regular trees X. Lisa Carbone provided many helpful discussions, and she proved the fundamental existence theorem for non-uniform lattices on uniform trees (Theorem (8.1)), which we had conjectured in the first draft of this book ([C1, 3]). Similarly, Gabe Rosenberg and Lisa Carbone, with help from the first author, have proved in general the Lattice Existence Theorem, originally only conjectured and proved in special cases here. We are grateful for their permission to publish this here as Appendix [BCR]. We are also indebted to Gabe Rosenberg for several new results cited below, and for his critical reading of the manuscript.

   The authors gratefully acknowledge long-term support and hospitality from the following institutions: the US National Science Foundation, the US–Israel Bi-national Science Foundation, Columbia University, the Landau Institute and the Institute for Advanced Studies of the Hebrew University of Jerusalem, and the University of Michigan.

We now proceed to a survey of the results and questions on tree lattices treated here and elsewhere.

0.2   *X*-lattices and *H*-lattices

To describe our results, and compare them with the Lie group case, we fix the following notation in this chapter.

(1)

$$X = \text{a locally finite tree.}$$

$$G = \text{Aut}(X), \text{ a locally compact group.}$$

$$H \leq G \text{ is a closed subgroup.}$$

$$\mu = \text{left Haar measure on } H \text{ with quotient } p : X \to H\backslash X.$$

$$G_H = \{g \epsilon G | gx \epsilon Hx \ \forall x \epsilon X\} = \{g \epsilon G | p \circ g = p\}.$$

By the **Lie group case** we understand that H is a simple rank 1 Lie group over a local non-archimedean field F, and X is the Bruhat–Tits tree of H. In this case lattices are understood to be subgroups of H.

In general, if

(U) $\qquad\qquad\qquad\qquad\qquad$ H $\quad$ is unimodular,

then $\mu(H_x)$ is constant on H-orbits, so we can define

(2) $$\mu(H\backslash\backslash X) := \sum_{x \in H\backslash VX} 1/\mu(H_x).$$

If $\Gamma \leq H$ is a discrete subgroup, then we have ((3.6) (2))

(3) $$\mathrm{Vol}(\Gamma\backslash\backslash X) = \mu(\Gamma\backslash H)\mu(H\backslash\backslash X).$$

It follows that

(4) $\qquad\quad$ Γ is an X-lattice $\Longleftrightarrow$ $\begin{cases} \Gamma \text{ is an } H\text{-lattice, and} \\ \text{(FV)} \qquad \mu(H\backslash\backslash X) < \infty. \end{cases}$

and

(5) $\qquad\quad$ Γ is a uniform X-lattice $\Longleftrightarrow$ $\begin{cases} \Gamma \text{ is an } H\text{-lattice, and} \\ \text{(F)} \qquad H\backslash X \quad \text{is finite.} \end{cases}$

Recall that a locally compact group that contains a lattice must be unimodular ([Rag 1], Ch. I).

0.3 Near simplicity

In the Lie group case, H contains a normal subgroup H^+ with H/H^+ abelian, and $H^+/Z(H^+)$ simple.

Tits [Ti1] established a similar result when G acts minimally on X, fixing no end of X. If G^+ denotes the subgroup generated by all edge stabilizers, then (cf. (6.12)):

(1) $\qquad$ For $N \leq G$ normalized by G^+, either $N = \{1\}$ or $G^+ \leq N$; and

(2) $\qquad$ G/G^+ is a free product of cyclic groups of orders ∞ and 2.

For example, if $X = X_d$, the regular tree of degree d, then, for $d = 2$, $G = D_\infty \cong (\mathbf{Z}/2\mathbf{Z}) * (\mathbf{Z}/2\mathbf{Z})$, and $G^+ = \{1\}$, whereas for $d \geq 3$, $[G : G^+] = 2$.

In fact (cf. (6.12)) Tits proves his simplicity theorem for other subgroups of G, for example, those of the form G_H ((0.2) (1)).

0.4   The structure of tree lattices

The uniform case

(1)   A group Γ is isomorphic to a uniform tree lattice iff Γ is finitely generated and virtually free. (See [B3], §8, and the references therein.)

The non-uniform case

Let Γ be a non-uniform X-lattice.

(2) Γ is not finitely generated, and

(3) Γ contains arbitrarily large finite subgroups. (See (5.16).)

(4)   Let C denote the set of Γ-conjugacy classes of maximal infinite-locally-finite subgroups of Γ. Then

$$0 \leq |C| \leq 2^{\aleph_0},$$

and all such cardinals are possible on the d-regular tree $X = X_d, d \geq 3$. In the Lie group case ($\Gamma \leq H$) we have $0 < |C| < \aleph_0$. (See (4.13) and (4.16)).

(5) Γ may or may not be residually finite. In fact Γ may have profinite completion $\hat{\Gamma} = \{1\}$. Moreover these phenomena occur on $X = X_d, d \geq 3$. In the Lie group case, Γ is residually finite. (See (10.3).)

(6) Γ can be a simple group. However, in this case Γ is locally finite (i.e., finitely generated subgroups are finite) and X is a "parabolic tree," i.e., X has a unique end. (See (9.15).)

0.5   Existence of lattices

Lie group case (cf. (0.2)); *char(F)* = 0

(1)   All H-lattices are uniform (Tamagawa, [T]).

(2)   There exist arithmetic H-lattices (Borel-Harder, [BH]).

(3)   There exist uncountably many H-conjugacy classes of H-lattices, hence there exist non-arithmetic ones (Lubotzky, [L3]).

Lie group case; *char(F)* = *p* > 0

(4)   There exist arithmetic non-uniform H-lattices. There exist arithmetic uniform H-lattices iff H is of type (A) (Borel-Harder, [BH]).

(5)   There exist uncountably many conjugacy classes of uniform, and of non-uniform H-lattices, hence non-arithmetic ones of both types (Lubotzky, [L3]).

Uniform tree lattices

(6)   **Uniform Existence Theorem** ([BK], (4.20) and (4.6)). *The following conditions are equivalent, in which case we call X a "uniform tree."*

(a) *(U) G is unimodular; and (F) $G \backslash X$ is finite.*

(b) *There exists a uniform X-lattice.*

(c) *X is the universal cover of a finite connected graph.*

(d) *There is a uniform X-lattice Γ such that $\Gamma \backslash X = G \backslash X$.*

There are examples satisfying (F) in (a), but not (U) ([BK], (4.12)).

(7)   **Uniform Commensurability Theorem** ([BK], (4.15)). *If $\Gamma, \Gamma' \leq G_H$ (cf. (0.2)) are uniform X-lattices, then Γ' and $g \Gamma g^{-1}$ are commensurable for some $g \in G_H$.*

This implies (cf. [BK], (4.16)):

(8)   **Leighton's Finite Common Covering Theorem** ([Lei]). *If A and A' are connected finite graphs with a common covering, then they have a common finite covering.*

We call the tree X *rigid* if $G(= \text{Aut}(X))$ is discrete. Beyond finite trees, the simplest example is the linear (2-regular) tree $X = X_2$, with $G \cong D_\infty$ (dihedral group). More interesting examples can be found in [BK], (4.12)2, and in Appendix [BT]. A rigid tree need not admit any lattices, but:

(9)   If X is rigid, then all X-lattices are uniform. (See (3.5).)

(10)   **Lattice Existence Theorem** (Appendix [BCR]). *There exists an X-lattice $\Gamma \leq G_H$ iff*

(U) *H is unimodular, and*

(FV) $\mu(H \backslash\backslash X) < \infty$.

Theorem (10) was conjectured in an earlier draft of this book. A full proof appears below in Appendix [BCR]. The conditions (U) and (FV) are clearly necessary for the existence of Γ (cf. (0.2) (3)). From (6), we know that (U) + (F) ($G \backslash X$ is finite) implies the existence of uniform X-lattices. Moreover we have a combinatorial "bounded denominator" condition (BD) on H ((2.6), (17) and (18)) such that

(11) $(U) + (BD) \Leftrightarrow$ there exists a discrete $\Gamma \leq G_H$ with $\Gamma \backslash X = H \backslash X$. Then Γ is an X-lattice iff (FV) $\mu(H \backslash\backslash X) < \infty$, and Γ is a uniform X-lattice iff (F) $H \backslash X$ is finite. (See [BK], (4.6).)

Many cases of Theorem (10) are proved in Chapter 7 below. In particular ((7.4)), when $\pi_1(H \backslash X)$ is finitely generated.

(12)   **Non-uniform lattices on uniform trees**

Let X be a uniform tree (cf. (6)). Does X admit a non-uniform lattice? The answer is negative in general, e.g., when X is rigid (cf. (9)). More generally, if X_0 is the minimal G-invariant subtree (which exists by (5.7), (5.11), and (9.7)), then rigidity of X_0 implies that all X-lattices are uniform, even though X itself need not be rigid. We call X *virtually rigid* if X_0 is rigid. We originally conjectured that this represents the only obstacle and we proved a number of special cases. This was then proved in complete generality by Lisa Carbone ([C1, 3]), who greatly generalized the techniques developed here.

(13)   **Theorem** (Carbone, [C1, 3]). *Suppose that*

> (U)   *H is unimodular;*
> (F)   *$H \backslash X$ is finite;*
> (ND)   *$G_H | X_0$ is not discrete.*

Then there is a non-uniform X-lattice $\Gamma \leq G_H$.

Many cases of (13) are proved below in Chapter 8, for example whenever $\pi_1(H \backslash X)$ is finitely generated, in particular when $H \backslash X$ is a tree. The full proof, which is long and intricate, can be found in [Ca1, 2].

0.6   The structure of $A = \Gamma \backslash X$

Let Γ be an X-lattice and $A = \Gamma \backslash X$.

Lie group case (Raghunathan [Rag 2], Lubotzky [L1-4]).

(1)   A is obtained from a finite graph A' by attaching finitely many infinite rays ("cusps"). In particular, $\pi_1(A) = \pi_1(A')$ is finitely generated.

General case. In contrast:

(2)   **"All Quotients Theorem"** ((4.17)). *Given any connected locally finite graph A, there exists a locally finite tree X, and an X-lattice Γ such that $\Gamma \backslash X \cong A$.*

Of course the X in (2) may not be regular, or even of bounded degree (even when A is of bounded degree). Nonetheless:

(3)   *Let $X = X_d$ ($d \geq 3$). Given $r, 0 \leq r \leq \aleph_0$, there is a non-uniform X-lattice Γ such that $\pi_1(A)$ is free on r generators. (See (4.11), Example 2.)*

(4)   *Let $X = X_d$ ($d \geq 3$). Every "combinatorially allowable geometric cusp structure" occurs on $A = \Gamma \backslash X$ for suitable X-lattices Γ. (See (4.13) and (4.14).)*

0.7 Volumes

Lie group case (Raghunathan [Rag 2], Lubotzky [L2])). Let

$$V(H) = \{\text{Vol}(\Gamma \backslash\backslash X) | \Gamma \text{ is an } X\text{-lattice}, \Gamma \leq H\} \subset (0, \infty).$$

Then

(1)   V(H) is a closed discrete subset of **R**. Let $V_{\min}$ = the least element of V(H).

For $H = SL_2(F)$, $V_{\min}$ has been determined by Lubotzky [L2] when $char(F) > 0$, and by Lubotzky and Weigel [LW] when $char(F) = 0$. When $F = \mathbf{F}_q((t^{-1}))$, $V_{\min}$ is achieved by $\Gamma = SL_2(\mathbf{F}_q[t])$.

Uniform tree lattice volumes. Put

$$d = \sup_{x \in VX} \deg(x) \qquad (< \infty \text{ when } X \text{ is uniform}).$$

Call an integer $D > 0$ a *d-number* if, for all primes p,

$$p | D \implies \begin{cases} p \leq d, \text{ and} \\ p = d \implies p^2 \nmid D. \end{cases}$$

It is easily verified that,

(2)   If Γ is a uniform X-lattice then $\text{Vol}(\Gamma \backslash\backslash X)$ is a rational number whose denominator is a d-number.

In fact, Inga Levich [Le] has shown that, on the d regular tree, there is no other restriction. More generally, Gabe Rosenberg has shown:

(3)   **Biregular uniform volumes** ([R]). Let $X = X_{d,d'}$, the (d, d') biregular tree, with $d \geq d' \geq 2$. Let $v = N/D$ be a positive rational number in reduced form. Then $v = \text{Vol}(\Gamma \backslash\backslash X)$ for some uniform X-lattice Γ iff D is a d-number and, if $d + d'$ is prime, then $(d + d') | N$.

This of course shows that such volumes $\text{Vol}(\Gamma \backslash\backslash X_{d,d'})$, for $d \geq 3$, are not bounded away from 0. This phenomenon follows also from the next result.

(4)   **Towers of lattices** (see [BK], §7, [R], and [CR1]).

(a)   If X admits a uniform lattice and is not rigid (cf. (0.5)(12)), then X admits a tower

$$\Gamma_1 < \Gamma_2 < \Gamma_3 < \cdots$$

of uniform lattices. (Note here that X may be virtually rigid.)

(b)   If X admits a non-uniform lattice (e.g., as in (a), by (13)) and if X is not parabolic (cf. (0.4)(6)), then X admits a tower

$$\Gamma_1 < \Gamma_2 < \Gamma_3 < \cdots$$

of non-uniform lattices.

At the same time,

(5)   **Overlattices.** If Γ is a uniform X-lattice and $m \geq 1$, then

$$U_\Gamma(m) := \{\Gamma' | \Gamma \leq \Gamma' \leq G, \quad [\Gamma' : \Gamma] \leq m\}$$

is finite. (See [BK], (6.5).)

From this and the Uniform Commensurability Theorem ((0.5)(8)), Lubotzky deduced ([L4]):

(6)   *There are only countably many G_H-conjugacy classes of uniform X-lattices*
        $\Gamma \leq G_H$.

On the other hand, he showed ([L3])) that this fails in the Lie group case for uniform H-lattices (cf. (0.5) (3) and (5)). Of course, (5) invites the

(7)   **Question.** *What can one say about the asymptotics of $|U(m)|$ as $m \to \infty$?*

This question seems to confront some difficult questions in finite group theory (cf. [BK], §7). One might think of this question as "dual" to the study of "subgroup growth," where one studies the asymptotics of the numbers of subgroups (in place of overgroups) of Γ of index $\leq m$ ([L5]).

Non-uniform tree lattice volumes

(8)   **Arbitrary Real Volumes Theorem** ((4.3)). *Let $X = X_d$, $d \geq 3$. Given*
        $v > 0$, there is an X-lattice Γ with $\mathrm{Vol}(\Gamma \backslash\backslash X) = v$ and $\Gamma \backslash X$ is a ray.

     G. Rosenberg [R] has generalized the arbitrary volume assertion of (8) to all uniform trees that admit a non-uniform lattice (cf. (0.5)(13)).

0.8   Centralizers, normalizers, commensurators

For subgroups $U, V \leq W$, we call U and V commensurable, denoted $U =_{virt} V$, if $U \cap V$ has finite index in U and in V. We also write

$$Z_V(U) = \{v \epsilon V \mid vu = uv \;\; \forall\, u \epsilon U\} \qquad \text{(centralizer)},$$
$$N_V(U) = \{v \epsilon V \mid vUv^{-1} = U\} \qquad \text{(normalizer)},$$
$$C_V(U) = \{v \epsilon V \mid vUv^{-1} =_{virt} U\} \qquad \text{(commensurator)}.$$

Thus we have an exact sequence

$$1 \to Z_V(U) \to N_V(U) \xrightarrow{ad_U} \mathrm{Aut}(U), \quad \text{and} \quad N_V(U) \leq C_V(U).$$

Lie group case

   If $\Gamma \leq H$ is an H-lattice, one has that

(1) $Z_H(\Gamma)$ is finite,

(2) $N_H(\Gamma)/\Gamma$ is finite, and

(3)   **Margulis Alternative** ([Mar]) Either $C_H(\Gamma)$ is discrete, and Γ is not arithmetic, or $C_H(\Gamma)$ is dense in H, in which case Γ is arithmetic (over some global field K, and $C_H(\Gamma)$ is essentially the group of K-rational points of H).

Uniform tree lattices

   Let Γ be a uniform X-lattice

(4) $Z_G(\Gamma)$ is finite, unless Γ is virtually cyclic (and X is "virtually linear") ([BK], (6.1)).

(5) $N_G(\Gamma)/\Gamma$ is finite ([BK], (6.4)).

(6)   **Uniform Density Theorem** ([Li1]). $C_G(\Gamma)$ *is dense in G.*

In view of the Uniform Commensurability Theorem ((0.5)(7)), $C_G(\Gamma) = C(X)$ is, up to conjugacy, independent of the uniform X-lattice Γ. In light of the Margulis alternative ((3) above), (5) suggests that we can think of uniform X-lattices as being "arithmetic," and of $C(X)$ as the "group of (global) rational points of G." This permitted Lubotzky to formulate a Congruence Subgroup Theorem, which, for regular trees, has been proved by Mozes [M2].

(7)   **Question.** *If Γ, Γ' are uniform X-lattices and $C_G(\Gamma) = C_G(\Gamma')$, does it follow that Γ and Γ' are commensurable?*

General tree lattices

   Let Γ be an X-lattice. Then Γ is finite iff X is finite. We assume now that Γ is infinite.

(8) $Z_G(\Gamma)$ is a closed subgroup of G. If Γ fixes no end of X then $Z_G(\Gamma)$ is a (direct) product of finite groups ((6.7)).

(9)   **Borel Density.** Γ acts minimally on X iff G acts minimally on X, in which case $Z_G(\Gamma) = \{1\}$ ((5.12) and (6.5)).

(10) $N_G(\Gamma)$ is a closed subgroup of G, and $N_G(\Gamma)/\Gamma$ is naturally a profinite group ((6.8)).

(11)   Suppose that $\Gamma \leq H = G_H$ and $\overline{C_H(\Gamma)} = H$ ($C_H(\Gamma)$ is dense in H). If Γ acts minimally on X then Γ is residually finite. In general (without minimality), there is an exact sequence

$$1 \to F \to \Gamma \to R \to 1$$

with F finite and R residually finite ((6.14) and (6.15)).

We next discuss an interesting class of examples, where we have made fairly detailed calculations. The reference numbers "(10.n)" refer to their location in Chapter 10, where details may be found.

Lattices of Nagao type ((10.1))

These are (non-uniform) X-lattices Γ which act with fundamental domain a ray,

$$x_0 \qquad x_1 \qquad x_2 \quad x_{n-1} \qquad x_n$$

and with stabilizers as follows:

– For $n \geq 1$, the stabilizers $\Gamma_n = \Gamma_{x_n}$ are increasing:

$$\Gamma_1 < \Gamma_2 < \Gamma_3 < \cdots .$$

– Put $\Gamma_0^+ = \Gamma_{x_0}$, $\Gamma_0 = \Gamma_0^+ \cap \Gamma_1$, and $\Gamma_\infty = \bigcup_{n \geq 0} \Gamma_n$. Then

$$\Gamma = \Gamma_0^+ *_{\Gamma_0} \Gamma_\infty.$$

– $\Gamma_0 < \Gamma_0^+$. This makes Γ act minimally on X, and $Z_G(\Gamma) = \{1\}$.

For such Γ we show that

$$N_G(\Gamma)/\Gamma \cong [\operatorname{Aut}(\Gamma_0^+; \Gamma_0) \times_{\operatorname{Aut}(\Gamma_0)} \operatorname{Aut}(\Gamma_\infty; (\Gamma_n)_{n \geq 0})]/ad_\Gamma(\Gamma_0).$$

(See (10.1) (26) for the notation.)

Example (10.2). Let $q = p^d$, p prime, $A = \mathbf{F}_q[t]$, a polynomial algebra, $F = \mathbf{F}_q(t)$, and $\hat{F} = \mathbf{F}_q((t^{-1}))$. Let $H = PSL_2$. The Bruhat–Tits tree of $H(\hat{F})$ is $X = X_{q+1}$, and $\Gamma = H(A)$ is an X-lattice as above, with $\Gamma_0^+ = H(\mathbf{F}_q)$, and

$$\Gamma_n = \left\{ \begin{bmatrix} a & b \\ 0 & a^{-1} \end{bmatrix} \,|\, a \epsilon \mathbf{F}_q^\times, \ \deg_t(b) \leq n \right\}, \ \text{ so that}$$

$$\Gamma_\infty \cong A \rtimes (\mathbf{F}_q^\times/\{\pm 1\}).$$

The decomposition $\Gamma = \Gamma_0^+ *_{\Gamma_0} \Gamma_\infty$ here was first proved by Nagao (cf. [Se], (II, 1.6)), whence our terminology. In $H(\hat{F})$ we have

$$N_{H(\hat{F})}(\Gamma) = \Gamma,$$
$$C_{H(\hat{F})}(\Gamma) = H(F), \ \text{ and}$$
$$\Gamma \text{ is residually finite.}$$

On the other hand ((10.2) (17))

$$N_G(\Gamma)/\Gamma \cong \left\{ \begin{array}{c} \mathbf{F}_q - \text{linear automorphisms } \alpha \text{ of } A \text{ such that} \\ \alpha(1) = 1 \text{ and } \deg_t(\alpha(t^n)) = n \ \forall n \end{array} \right\},$$

an infinite pro-p-group. S. Mozes [M2] has recently shown that $C_G(\Gamma)$ is dense in G.

Example (10.3). Let E be the unramified extension of $\mathbf{Q}_p$ with residue field $\mathbf{F}_q = R/pR$, R the integral closure of $\mathbf{Z}_p$ in E. On $X = X_{q+1}$ there is a lattice Γ of Nagao type with $\Gamma_0^+ = H(\mathbf{F}_q)$, as above, but $\Gamma_\infty \cong (E/R){\rtimes}(\mathbf{F}_q^\times/\{\pm1\})$ (in place of $A{\rtimes}(\mathbf{F}_q^\times/\{\pm1\})$) in (10.2) above). In this case we have the profinite completion $\hat{\Gamma} = \{1\}$, and so $C_G(\Gamma) = N_G(\Gamma)$. Moreover, writing $\mathbf{Z}_p$ for p-adic integers,

$$N_G(\Gamma)/\Gamma \cong (1 + pR) \text{ (multiplicative group)}$$
$$\cong \begin{cases} \mathbf{Z}_p^d & \text{if } p > 2, \\ \mathbf{Z}_2^d \oplus \mathbf{F}_2 & \text{if } p = 2. \end{cases}$$

This $C_G(\Gamma)$ is not dense in G.

Example (10.4). In (10.4) one finds examples, based on a construction supplied by Peter Neumann (Appendix [PN]), that give the first examples of non-uniform Γ on regular trees with $N_G(\Gamma)/\Gamma$ finite. Here Γ is of Nagao type, and, for a suitable integer $q > 1$,

$$\deg(x_n) = \begin{cases} q+1 & \text{for } n \geq 1, \\ q_0 + 1 & \text{for } n = 0. \end{cases}$$

For $q_0 = q - 1$ (so X is not regular), our example is self-normalizing,

$$N_G(\Gamma) = \Gamma.$$

With $q_0 = q$ (so $X = X_{q+1}$ is regular) we only have

$$N_G(\Gamma)/\Gamma \text{ is finite.}$$

However $N = N_G(\Gamma)$ is still of Nagao type (in fact $N\backslash X = \Gamma\backslash X$), and Γ can be chosen so that $N_G(N) = N$.

Example (10.5). The X in Example (10.5) has unbounded degree, Γ_∞ is the infinite symmetric group, Γ contains an index 2 subgroup Γ' with Γ'_∞ the (simple) infinite alternating group, $\hat{\Gamma}' = \{1\}$, and

$$C_G(\Gamma) = N_G(\Gamma') = C_G(\Gamma') = \Gamma,$$

which is not dense in G.

Example (6.21) (3) (Burger–Mozes [BM1]). If C_m denotes a cyclic group of order m, then in this example,

$$\Gamma_n = (C_2)^n \quad \text{for } n \geq 0, \text{ and}$$
$$\Gamma_0^+ = C_b, \quad \text{with } b > 3.$$

(Thus X is not regular.) *In this example $C_G(\Gamma)$ is dense in G.*

Examples (10.2) and (6.21) are the only cases of non-uniform tree lattices that have been shown to have dense commonsurators. We show in (6.15) that if an X-lattice Γ acts minimally on X and $C_G(\Gamma)$ is dense in G, then Γ must be residually finite. This explains why examples like (10.3) above have non-dense commensurators.

Chapter 1

Lattices and Volumes

We recall some basic facts about lattices in locally compact groups. A good reference is [Rag 1], Ch. 1.

1.1 Haar measure

We fix a locally compact group H with a left Haar measure μ. For measurable $U \subset H$ and $g \in H$ we have

$$\mu(gU) = \mu(U), \quad \text{and} \quad \mu(Ug) = \mu(U)\Delta(g),$$

where $\Delta : H \to \mathbf{R}^\times$ is the *modular character* of H. One calls H *unimodular* if $\Delta = 1$. When H is discrete, we choose μ so that $\mu(\{g\}) - 1$ for $g \in H$

1.2 Lattices and unimodularity

Let $\Gamma \le H$ be a discrete subgroup, with quotient $p : H \to \Gamma \backslash H$. Then $\Gamma \backslash H$ carries a measure μ (or $\mu_{\Gamma \backslash H}$) so that p is locally measure preserving. The measure $\mu_{\Gamma \backslash H}$, like μ, is "Δ-semi-invariant" for right translation;

$$(1) \qquad\qquad \mu_{\Gamma \backslash H}(Vg) = \mu_{\Gamma \backslash H}(V)\Delta(g)$$

for measurable $V \subset \Gamma \backslash H$. In case $\mu_{\Gamma \backslash H}(\Gamma \backslash H) < \infty$, then, taking $V = \Gamma \backslash H$ in (1) we see that $\Delta = 1$ (H is *unimodular*). In this case we call Γ an H-*lattice*. We call Γ a *uniform* H-*lattice* if $\Gamma \backslash H$ is compact.

1.3 Compact open subgroups

Let $KO(H)$ denote the set of compact open subgroups of H. Any two members of $KO(H)$ are commensurable. A function $\lambda : KO(H) \to \mathbf{C}$ is called *multiplicative* if, whenever $K' \leq K$ are in $KO(H)$, we have $\lambda(K) = [K : K']\lambda(K')$. For example, $K \mapsto \mu(K)$ is multiplicative. If λ is multiplicative, $K_0, K_1, K' \in KO(H)$, and $K' \leq K_0 \cap K_1$, it follows that $\lambda(K_1) = \frac{[K_1:K']}{[K_0:K']}\lambda(K_0)$; thus $\lambda(K_0)$ determines $\lambda(K_1)$ for all $K_1 \in KO(H)$. It follows that $\lambda(K) = c\mu(K)$ for all $K \in KO(H)$, where $c = \lambda(K_0)/\mu(K_0)$.

1.4   Proposition. *Let $\Gamma \leq H$ be a discrete subgroup, $Y = \Gamma \backslash H$, and $y \in Y$. For $K \leq H$ a compact subgroup we have that K_y is finite, and*

$$\mu_{\Gamma \backslash H}(y \cdot K) \cdot |K_y| = \mu(K).$$

Proof. The stabilizers H_y are conjugates of Γ, hence discrete, so K_y is discrete and compact, i.e., finite. Since $y \cdot K \cong K_y \backslash K$, and $p : H \to \Gamma \backslash H = Y$ is locally measure preserving it follows that $\mu_{\Gamma \backslash H}(y \cdot K) = \mu(K_y \backslash K) = \mu(K)/|K_y|$.

1.5   Discrete group covolumes

Let X be a left *H-set with compact open stabilizers H_x ($x \in X$). Let $\Gamma \leq H$ be a discrete subgroup.* Then the stabilizers Γ_x are finite, and we define

$$(1) \qquad\qquad \mathrm{Vol}(\Gamma \backslash\backslash X) = \sum_{x \in \Gamma \backslash X} 1/|\Gamma_x|.$$

Here one can interpret "$\Gamma \backslash\backslash X$" as $\Gamma \backslash X$, with the "weight" $1/|\Gamma_x|$ attached to the image of x. Consider the commutative diagram of natural projections

$$(2) \qquad\qquad \begin{array}{ccc} & X & \\ p_\Gamma \downarrow & \searrow^{p_H} & \\ \Gamma \backslash X & \xrightarrow{\ p\ } & H \backslash X \end{array}$$

For $x \in X$ and $z = p_H(x)$ we have

$$(3) \qquad\qquad p^{-1}(z) = \Gamma \backslash (H \cdot x) \cong \Gamma \backslash H / H_x.$$

Let $S \subset H$ be a set of representatives of $\Gamma \backslash H / H_x$:

$$(4) \qquad\qquad \begin{aligned} H &= \textstyle\coprod_{s \in S} \Gamma \cdot s \cdot H_x, \quad \text{and} \\ \Gamma \backslash H &= \textstyle\coprod_{s \in S} \Gamma \backslash (\Gamma \cdot s \cdot H_x). \end{aligned}$$

Thus $\{sx | s \in S\} \subset X$ represents $\Gamma \backslash (H \cdot x)$, so

$$(5) \qquad\qquad \mathrm{Vol}(\Gamma \backslash\backslash (H \cdot x)) = \sum_{s \in S} 1/|\Gamma_{sx}|.$$

Putting $Y = \Gamma \backslash H$ and $y_0 = \Gamma \cdot 1 \in Y$, we have $\Gamma \backslash (\Gamma \cdot s \cdot H_x) = y_0 s \cdot H_x$. From (4) we then have

$$\mu_{\Gamma \backslash H}(\Gamma \backslash H) = \sum_{s \in S} \mu_{\Gamma \backslash H}(y_0 s \cdot H_x)$$

$$\overset{(1.4)}{=} \sum_{s \in S} \mu(H_x)/|(H_x)_{y_0 s}|.$$

Now $(H_x)_{y_0 s} = H_x \cap s^{-1} \Gamma s$ is conjugate to $s H_x s^{-1} \cap \Gamma = H_{sx} \cap \Gamma = \Gamma_{sx}$. Thus, continuing from above,

$$\mu_{\Gamma \backslash H}(\Gamma \backslash H) = \sum_{s \in S} \mu(H_x)/|\Gamma_{sx}|$$

$$= \mu(H_x) \sum_{s \in S} 1/|\Gamma_{sx}|$$

$$\overset{(5)}{=} \mu(H_x) \mathrm{Vol}(\Gamma \backslash\backslash (H \cdot x)), \quad \text{i.e.,}$$

(6) $$\mu_{\Gamma \backslash H}(\Gamma \backslash H) = \mu(H_x) \cdot \mathrm{Vol}(\Gamma \backslash\backslash (H \cdot x)), \quad \forall\, x \in X.$$

Thus,

(7)   The following conditions are equivalent:

   (a) $\mathrm{Vol}(\Gamma \backslash\backslash (H \cdot x)) < \infty$ for some $x \in X$.

   (a') $\mathrm{Vol}(\Gamma \backslash\backslash (H \cdot x)) < \infty$ for every $x \in X$.

   (b) Γ is an H-lattice.

In a similar fashion we see that

(8)   The following conditions are equivalent:

   (a) Some fiber $p^{-1}(p_H(x))$ ($\cong \Gamma \backslash H / H_x$) is finite.

   (a') Every fiber of p is finite.

   (b) Γ is a uniform H-lattice.

If H is discrete, then the conditions of (7) imply those of (8), and then (6) takes the form

(6') $$[H : \Gamma] = |H_x| \cdot \mathrm{Vol}(\Gamma \backslash\backslash (H \cdot x)) < \infty.$$

In fact, the stabilizers Γ_y ($y \in (H \cdot x)$) then have orders $\leq |H_x|$, so the finiteness of $\mathrm{Vol}(\Gamma \backslash\backslash (H \cdot x))$ implies the finiteness of $\Gamma \backslash (H \cdot x) \cong \Gamma \backslash H / H_x$, and so also of $\Gamma \backslash H$. When H is unimodular, $\mu(H_x)$ ($x \in X$) is constant on H-orbits, so we can define

(9) $$\mu(H \backslash\backslash X) := \sum_{x \in H \backslash X} 1/\mu(H_x).$$

Then we have

$$\begin{aligned}
\mathrm{Vol}(\Gamma \backslash\backslash X) &= \sum_{x \in H \backslash X} \mathrm{Vol}(\Gamma \backslash\backslash H \cdot x) \\
&\overset{(6)}{=} \sum_{x \in H \backslash X} \mu_{\Gamma \backslash H}(\Gamma \backslash H) / \mu(H_x) \\
&= \mu_{\Gamma \backslash H}(\Gamma \backslash H) \cdot \mu(H \backslash\backslash X), \quad \text{i.e.,}
\end{aligned}$$

$$(10) \qquad \mathrm{Vol}(\Gamma \backslash\backslash X) = \mu_{\Gamma \backslash H}(\Gamma \backslash H) \cdot \mu(H \backslash\backslash X).$$

1.6   **Corollary.** *Let X be a left H-set with compact open stabilizers and $\Gamma \leq H$ a discrete subgroup. The following conditions are equivalent:*

(a) $\mathrm{Vol}(\Gamma \backslash\backslash X) < \infty$.

(b) (i) Γ *is an H-lattice (hence H is unimodular), and*

   (ii) $\mu(H \backslash\backslash X) < \infty$.

In this case, we have

$$\mathrm{Vol}(\Gamma \backslash\backslash X) = \mu_{\Gamma \backslash H}(\Gamma \backslash H) \cdot \mu(H \backslash\backslash X).$$

In case H itself is discrete, and μ gives measure 1 to each point, (1.6) takes the form:

1.7   **Corollary.** *Let X be a left H-set with finite stabilizers, and $\Gamma \leq H$. The following conditions are equivalent.*

(a) $\mathrm{Vol}(\Gamma \backslash\backslash X) < \infty$.

(b) (i) $[H : \Gamma] < \infty$, *and*

   (ii) $\mathrm{Vol}(H \backslash\backslash X) < \infty$.

In this case, the fibers of $p : \Gamma \backslash X \to H \backslash X$ are finite (see (1.5)(18)), and we have

$$\mathrm{Vol}(\Gamma \backslash\backslash X) = [H : \Gamma]\mathrm{Vol}(H \backslash\backslash X).$$

Chapter 2

Graphs of Groups and Edge-Indexed Graphs

We review here some basic material that can be found, for example, in [B3]. (See also [Se] and [BK].)

2.1 Graphs

A graph A consists of sets VA of vertices, EA of oriented edges, end point maps $\partial_0, \partial_1 : EA \to VA$, and orientation reversal $e \mapsto \bar{e}$ $(e \epsilon EA)$, satisfying $\partial_i \bar{e} = \partial_{1-i} e$, and $\bar{\bar{e}} = e \neq \bar{e}$. For $a \epsilon VA$ we put $E_0(a) = \partial_0^{-1}(a)$, and $\deg(a) = |E_0(a)|$. We call A *locally finite* if $\deg(a) < \infty$ for all a, and d *-regular* if $\deg(a) = d$ for all a.

A *path* of length n in A is a sequence $\gamma = (e_1, \ldots, e_n)$ of edges such that $\partial_1 e_i = a_i = \partial_0 e_{i+1}$, $1 \leq i < n$. Putting $a_0 = \partial_0 e_1$ and $a_n = \partial_1 e_n$, we call γ a path from a_0 to a_n. The (empty) path of length 0 from a_0 to a_0 is denoted (a_0). We call γ *closed* if $a_0 = a_n$. We call γ *reduced* if it contains no reversals, $(e_i, e_{i+1}) = (e_i, \bar{e}_i)$. We call A *connected* if $A \neq \phi$ and any two vertices can be joined by a path, and we call A a *tree* if A is connected and contains no closed reduced paths of length $n > 0$.

Let A be a connected graph. An edge $e \epsilon EA$ is called *separating* if the removal of $\{e, \bar{e}\}$ disconnects A. The resulting two connected graphs are denoted $A_0(e)$ and $A_1(e)$, where $A_i(e)$ is the component containing $\partial_i e$ $(i = 0, 1)$.

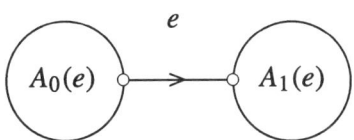

2.2   Morphisms and actions

A *morphism* $s : A \to B$ of graphs maps VA to VB, EA to EB, and satisfies $s(\partial_i e) = \partial_i(se)$, $s(\bar{e}) = \overline{s(e)}$, for $e \in EA$, $i = 0, 1$. If $a \in VA$ and $s(a) = b$, then we have the local map $s_{(a)} : E_0(a) \to E_0(b)$. We say s has a property (e.g., surjectivity, bijectivity, ...) locally if each of the maps $s_{(a)}$ has this property.

Suppose that a group Γ acts on a graph A. We call $s \in \Gamma$ an *inversion* on A if $se = \bar{e}$ for some $e \in EA$. If Γ acts without inversions, then we can form the quotient graph, and projection,

$$p : A \to \Gamma \backslash A.$$

This is locally surjective. In fact, if $a \in VA$ and $p(a) = b$, then $p_{(a)} : E_0(a) \to E_0(b)$ is the quotient map modulo the action of the stabilizer Γ_a. Hence, if $e \in E_0(a)$ and $p(e) = f$, then

$$[\Gamma_a : \Gamma_e] = |p_{(a)}^{-1}(f)| = |\Gamma_a \cdot e|,$$

which depends only on f and the projection p.

2.3   Graphs of groups

A *graph of groups* $\mathbf{A} = (A, \mathcal{A})$ consists of a connected graph A, groups $\mathcal{A}_a$ $(a \in VA)$ and $\mathcal{A}_e = \mathcal{A}_{\bar{e}}$ $(e \in EA)$, and monomorphisms $\alpha_e : \mathcal{A}_e \to \mathcal{A}_{\partial_0 e}$ $(e \in EA)$.

Choosing a base point $a_0 \in VA$, one can define a *fundamental group* $\Gamma = \pi_1(\mathbf{A}, a_0)$, a *universal covering tree* $X = \widetilde{(\mathbf{A}, a_0)}$, and an action without inversion of Γ on X with a morphism $p : X \to A$ which can be indentified with the quotient projection modulo $\Gamma : A = \Gamma \backslash X$. Further, if $e \in EX$, $x = \partial_0 e$, $p(x) = a$, and $p(e) = f$, then $\alpha_f : \mathcal{A}_f \to \mathcal{A}_a$ is isomorphic to the inclusion $\Gamma_e \to \Gamma_x$.

We call the graph of groups $\mathbf{A}$ *faithful* if the action of Γ on X is faithful, i.e., if the action defining homomorphism $\Gamma \to \text{Aut}(X)$ is injective. In general any $\mathbf{A} = (A, \mathcal{A})$ has a *faithful quotient* $\mathbf{A}' = (A, \mathcal{A}')$ with groups $\mathcal{A}'_a (a \in VA)$ and $\mathcal{A}'_e$ $(e \in EA)$ which are quotients of $\mathcal{A}_a$ and $\mathcal{A}_e$, respectively, and so that the diagrams, for $\partial_0 e = a$,

$$
\begin{array}{ccc}
\mathcal{A}_e & \xrightarrow{\alpha_e} & \mathcal{A}_a \\
\downarrow & & \downarrow \\
\mathcal{A}'_e & \xrightarrow{\alpha'_e} & \mathcal{A}'_a
\end{array}
$$

commute and induce bijections $\mathcal{A}_a/\alpha_e \mathcal{A}_e \to \mathcal{A}'_a/\alpha'_a \mathcal{A}'_e$. We then have $\widetilde{(\mathbf{A}', a_0)} = X = \widetilde{(\mathbf{A}, a_0)}$, and

$$\Gamma' = \pi_1(\mathbf{A}', a_0) = Im(\Gamma = \pi_1(\mathbf{A}, a_0) \to \text{Aut}(X)).$$

When $\mathbf{A}$ is a *graph of finite groups* $\mathcal{A}_a$ we define the *volume* of $\mathbf{A}$ by

$$\mathrm{Vol}(\mathbf{A}) = \sum_{a \epsilon VA} \frac{1}{|\mathcal{A}_a|}.$$

2.4   Quotient graphs of groups

Let X be a connected graph on which a group Γ acts without inversion, and with quotient

$$p : X \to A = \Gamma\backslash X.$$

Then one can construct a *quotient graph of groups*

$$\mathbf{A} = (A, \mathcal{A}) = \Gamma\backslash\backslash X$$

so that, for $e \epsilon EX$, $x = \partial_0 e$, $p(x) = a$, and $p(e) = f$, $\alpha_f : \mathcal{A}_f \to \mathcal{A}_a$ is isomorphic to the *inclusion* $\Gamma_e \to \Gamma_x$. In particular,

(1) $$[\mathcal{A}_a : \alpha_f \mathcal{A}_f] = [\Gamma_x : \Gamma_e] = |\Gamma_x \cdot e| = |p_{(x)}^{-1}(f)|.$$

Let $\tilde{\Gamma} = \pi_1(\mathbf{A}, a_0)$ (for some $a_0 \epsilon VA$), acting on $\tilde{X} = \widetilde{(\mathbf{A}, a_0)}$. Then there is a homomorphism $f : \tilde{\Gamma} \to \Gamma$, and an f-equivariant morphism $q : \tilde{X} \to X$ so that $p \circ q : \tilde{X} \to A = \tilde{\Gamma}\backslash\tilde{X}$ is the natural projection. Further f is an isomorphism iff q is an isomorphism, and iff X is a tree.

Thus: When a group Γ acts without inversion on a tree X, we can naturally identify Γ and X with the fundamental group and universal covering tree, respectively, of the quotient graph of groups, $\Gamma\backslash\backslash X$. Further there is an exact sequence

$$1 \to N \to \Gamma \to \pi_1(\Gamma\backslash X) \to 1,$$

where N is the (normal) subgroup of Γ generated by all stabilizers Γ_x ($x \epsilon VX$).

2.5   Edge-indexed graphs and their groupings

Let $\mathbf{A} = (A, \mathcal{A})$ be a graph of groups. For $e \epsilon EA$, $\partial_0 e = a$, put

(1) $$i(e) \ (= i_{\mathbf{A}}(e)) := [\mathcal{A}_a : \alpha_e \mathcal{A}_e].$$

Thus i assigns a cardinal $i(e) > 0$ to each $e \epsilon EA$. (We shall be concerned only with the case where all $i(e)$ are finite.) We call

(2) $$I(\mathbf{A}) := (A, i)$$

the *edge-indexed graph associated to* **A**. Further, given (A, i), we call **A** satisfying (1) a *grouping* of (A, i).

Let $a_0 \epsilon VA$, and put $X = \widetilde{(\mathbf{A}, a_0)}$. A fundamental observation (cf. [B3], (1.18)) of which we shall make extensive use is that:

(3) $X = \widetilde{(\mathbf{A}, a_0)}$, with its projection $p : X \to A$, depends, up to (non-canonical) isomorphism (over Id_A) only on (A, i, a_0) and not on the grouping; we therefore also denote it $X = \widetilde{(A, i, a_0)}$.

Recall from (2.4) (1) that p determines (A, i). In fact, if $x \epsilon VX$, $p(x) = a$, and $e \epsilon E_0(a)$, then we have the local map $p_{(x)} : E_0(x) \to E_0(a)$, and $i(e) = |p_{(x)}^{-1}(e)|$.

Given any edge-indexed graph (A, i), and $a_0 \epsilon VA$, it follows from (3) that each grouping **A** of (A, i) produces a group $\Gamma = \pi_1(\mathbf{A}, a_0)$ acting without inversions on X with quotient $p : X \to A = \Gamma \backslash X$. If we replace **A** by its faithful quotient **A**′, we obtain the image of Γ, $\Gamma' \leq G = \mathrm{Aut}(X)$, whose stabilizers are isomorphic to the groups $\mathcal{A}'_a$ and $\mathcal{A}'_e$ of **A**′.

Let (B, j) be another edge-indexed graph and $q : B \to A$ a graph morphism. We call

(4) $q : (B, j) \to (A, i)$ a *covering* of edge-indexed graphs

if $\forall b \epsilon VB, q(b) = a$, and $e \epsilon E_0(a)$, the local map $q_{(b)} : E_0(b) \to E_0(a)$ satisfies

(5) $$i(e) = \sum_{f \epsilon q_{(b)}^{-1}(e)} j(f).$$

In this case, we can identify

(6) $$\widetilde{(A, i, a)} = X = \widetilde{(B, j, b)}$$

so that the diagram of projections

(7)
$$
\begin{array}{ccc}
 & X & \\
{\scriptstyle p_B}\downarrow & & \searrow{\scriptstyle p_A} \\
B & \xrightarrow[q]{} & A
\end{array}
$$

commutes. This is easily verified, for example by the same considerations as in [B3], (1.18).

2.6   Unimodularity, volumes, bounded denominators

Let (A, i), be an edge-indexed graph. We always assume that A is connected, and we henceforth assume also that all indices $i(e)$ are finite:

$$i : EA \to \mathbf{Z}, \qquad i(e) > 0 \ \forall e.$$

We put

$$(1) \qquad \Delta(e) = \frac{i(\bar{e})}{i(e)} \quad \text{for } e \epsilon EA$$

and

$$(2) \qquad \Delta(\gamma) = \Delta(e_1)\ldots\Delta(e_n) \quad \text{for a path } \gamma = (e_1, \ldots, e_n) \text{ in } A.$$

Given a function $N : VA \to \mathbf{R}^{\times}$ we define N on edges by,

$$(3) \qquad N(e) = \frac{N(\partial_0 e)}{i(e)} \quad \text{for } e \epsilon EA.$$

(4)   The following conditions on (A, i) are equivalent:

(a) $\Delta(\gamma) = 1$ for all closed paths γ.

(b)   There is a function $N : VA \to \mathbf{R}^{\times}$ such that $N(e) = N(\bar{e}) \; \forall \, e \epsilon EA$.

Further, the function N in (b) is unique up to a constant factor. The condition $N(e) = N(\bar{e})$ in (b) can be rewritten as

$$(5) \qquad \Delta(e) = \frac{N(\partial_1 e)}{N(\partial_0 e)} \quad (= \Delta(\bar{e})^{-1}) \quad \forall \, e \epsilon EA.$$

This condition implies that

$$(6) \qquad \Delta(\gamma) = \frac{N(b)}{N(a)} \quad \text{for any path } \gamma \text{ from } a \text{ to } b,$$

whence (b) $\Rightarrow$ (a) in (4).

Conversely, assuming (4) (a), it follows that, if γ is a path from a to b in A, then $\Delta(\gamma)$ depends only on (a, b), not on γ. We put

$$(7) \qquad \frac{\Delta b}{\Delta a} := \Delta(\gamma) \quad \text{for any path } \gamma \text{ from } a \text{ to } b.$$

If $e \epsilon EA$, $\partial_0 e = b$, $\partial_1 e = c$, then

$$\left(\frac{\Delta c}{\Delta a}\right) \Big/ \left(\frac{\Delta b}{\Delta a}\right) = \frac{\Delta c}{\Delta b} = \Delta(e),$$

which is just condition (5) for the function N_a defined by

$$(8) \qquad N_a(b) := \frac{\Delta b}{\Delta a},$$

whence (a) $\Rightarrow$ (b) in (4).

When the conditions of (4) hold, we say that (A, i) *is unimodular.* In this case, it follows from (6) that, for any N as in (4)(b),

$$(9) \qquad \frac{N(b)}{N(a)} = \frac{\Delta b}{\Delta a} \quad \forall \, a, b \epsilon VA,$$

whence the uniqueness of N up to a constant factor:

$$(10) \qquad N = N(a) \cdot N_a.$$

We further define the *volume* of (A, i), at the base point $a \epsilon VA$, by

$$(11) \qquad \mathrm{Vol}_a((A, i)) = \sum_{b \epsilon VA} \frac{1}{N_a(b)} = \sum_{b \epsilon VA} \frac{\Delta a}{\Delta b}.$$

Note that

$$(12) \qquad \mathrm{Vol}_{a'}((A, i)) = \left(\frac{\Delta a'}{\Delta a} \right) \cdot \mathrm{Vol}_a((A, i)),$$

so it makes sense to speak of *finite volume,*

$$(13) \qquad \mathrm{Vol}((A, i)) < \infty,$$

without reference to a base point.

In case $(A, i) = I(\mathbf{A})$, where $\mathbf{A} = (A, \mathcal{A})$ is a graph of *finite* groups, then

$$(14) \qquad N(a) := |\mathcal{A}_a|$$

satisfies (4)(b), since, if $e \epsilon E_0(a)$, then (3) shows that

$$N(e) = |\mathcal{A}_e| = |\mathcal{A}_{\bar{e}}| = N(\bar{e}).$$

Further, $N = |\mathcal{A}_a| \cdot N_a$, so

$$(15) \qquad \mathrm{Vol}(\mathbf{A}) = \sum_{b \epsilon VA} \frac{1}{|\mathcal{A}_b|} = \frac{1}{|\mathcal{A}_a|} \mathrm{Vol}_a((A, i)).$$

Moreover, for $e \epsilon EA$, we have

$$(16) \qquad N_a(e) = \frac{N(e)}{|\mathcal{A}_a|} = \frac{|\mathcal{A}_e|}{|\mathcal{A}_a|},$$

and so,

$$(17) \qquad \{N_a(e) \epsilon \mathbf{Q}^\times | \ e \epsilon EA\} \ \text{has bounded denominators.}$$

More generally, without assuming that (A, i) is unimodular, we say that (A, i) *has bounded denominators* if for some (hence every) $a \epsilon VA$,

(18)
$$\left\{ \frac{\Delta(\gamma)}{i(e)} \epsilon \mathbf{Q}^\times \mid e \epsilon EA, \gamma \text{ a path from } a \text{ to } \partial_0 e \right\}$$

has bounded denominators.

This implies unimodularity. For suppose that γ is a closed path at a and $\Delta(\gamma) \neq 1$. Replacing γ by γ^{-1}, if necessary, we can assume that $\Delta(\gamma)$ has denominator > 1. Then, for $e \epsilon E_0(a)$, the numbers $\frac{\Delta(\gamma^n)}{i(e)} = \frac{\Delta(\gamma)^n}{i(e)}$ have unbounded denominators. When A is finite, unimodularity implies (hence is equivalent to) bounded denominators.

Assume that (A, i) has bounded denominators. Choose a denominator D so that $N := D \cdot N_a$ takes integer values

$$N(b), N(e) \epsilon \mathbf{Z} \quad \forall b \epsilon VA, \, e \epsilon EA.$$

Then it follows easily (see [BK]), that there is a finite (cyclic) grouping $\mathbf{A} = (A, \mathcal{A})$ of (A, i) such that

$$|\mathcal{A}_b| = N(b), |\mathcal{A}_e| = N(e) \quad \forall b \epsilon VA, \, e \epsilon EA.$$

This explains the following equivalence.

(19)   The following conditions on (A, i) are equivalent.

   (a) (A, i) has bounded denominators (hence is unimodular).
   (b) There is a function $N : VA \to \mathbf{Z} - \{0\}$ such that $N(e) = N(\bar{e}) \; \forall e \epsilon EA$ (cf. (3)).
   (c) There is a finite grouping $\mathbf{A}$ of (A, i).

If, in (19)(c), we replace $\mathbf{A}$ by its faithful quotient, then $\Gamma = \pi_1(\mathbf{A}, a_0)$ is a subgroup of $G = \mathrm{Aut}(X)$, $X = (\widetilde{A, i, a_0})$, and, for $e \epsilon EX$, $x = \partial_0 e$, $a = p(x)$, $f = p(e)$, (where $p : X \to A = \Gamma \backslash X$), the inclusion $\Gamma_e \leq \Gamma_x$ is isomorphic to $\alpha_f : \mathcal{A}_f \to \mathcal{A}_a$.

Chapter 3

Tree Lattices

3.1 Topology on $G = \mathrm{Aut}\,(X)$

Let X be a locally finite tree. Then $G = \mathrm{Aut}(X)$ is a locally compact group, where two automorphisms are close if they agree on a large finite subtree. For $x \in VX$ the stabilizer G_x is open, and compact; in fact

$$G_x = \varprojlim_r (G_x | B_x(r)) \qquad (r \to \infty),$$

where $B_x(r)$, the ball of radius r centered at x, is a finite subtree (by local finiteness), and so G_r is a profinite group. For $x, y \in VX$, G_x, and G_y are commensurable: If $d(x, y) = r$ then $G_x \cap G_y$ contains $Ker(G_x \xrightarrow{\mathrm{res}} \mathrm{Aut}(B_x(r)))$.

3.2 Tree lattices

A subgroup $\Gamma \leq G$ is *discrete* if Γ_x is finite for some (hence every) $x \in VX$. We then define the *volume*,

$$(1) \qquad\qquad \mathrm{Vol}(\Gamma \backslash\backslash X) := \sum_{x \in \Gamma \backslash VX} 1/|\Gamma_x|,$$

and call Γ an *X-lattice* if this is finite. We call Γ a *uniform X-lattice* if $\Gamma \backslash X$ is finite. If $x \in VX$ projects to $a \in \Gamma \backslash VX = V(\Gamma \backslash X)$, then it follows from (2.6)(11) that

$$(2) \qquad\qquad |\Gamma_x| \cdot \mathrm{Vol}(\Gamma \backslash\backslash X) = \mathrm{Vol}_a(I(\Gamma \backslash\backslash X)).$$

Thus:

(3) Γ is an *X*-lattice iff $\Gamma \backslash\backslash X$ is a graph of *finite* groups (discreteness) and $\mathrm{Vol}(I(\Gamma \backslash\backslash X)) < \infty$ (cf. (2.6)(13)).

(4) Γ is a uniform *X*-lattice iff $\Gamma \backslash\backslash X$ is a finite graph of finite groups.

3.3   The group G_H of deck transformations

Let $H \leq G$ be a subgroup without inversions. Let

$$p : X \to A = H \backslash X$$

be the quotient graph projection. Recall that

$$(A, i) = I(H \backslash\backslash X)$$

depends only on p. In fact, if $f \epsilon EX$, $\partial_0 f = x$, $p(f) = e$, and $p(x) = a = \partial_0 e$, then we have the local map

$$p_{(x)} : E_0(x) \to E_0(a),$$

and

(1) $$i(e) = |p_{(x)}^{-1}(e)| = [H_x : H_f] = |H_x \cdot f|.$$

Likewise the following group of deck transformations G_H depends only on p:

(2) $$G_H = \{g \epsilon G | \ p \circ g = p\} = G_{(A,i)}.$$

This is the group of automorphisms of X that preserve all H-orbits. It is a closed subgroup of G, hence contains the closure $\bar{H}$ of H,

(3) $$H \leq \bar{H} \leq G_H.$$

It follows from (1) that

(4) $$(A, i) = I(H \backslash\backslash X) = I(\bar{H} \backslash\backslash X) = I(G_H \backslash\backslash X).$$

For $x \epsilon VX$, $G_H \cdot x = H \cdot x$, so

(5) $$G_H = H \cdot (G_H)_x,$$

with $(G_H)_x$ compact (open). Hence,

(6)   If H is discrete, then H is a uniform G_H-lattice.

(7) H is called $(X\text{-})saturated$ if $H = G_H$.

Pick a base point $a \epsilon VA$, and identify X with $\widetilde{(A, i, a)}$. Then evidently,

(8)   If $\mathbf{A}$ is a faithful grouping of (A, i), and $\Gamma = \pi_1(\mathbf{A}, a)$, then $\Gamma \leq G_H$, and $I(\Gamma \backslash\backslash X) = (A, i)$.

More generally, suppose that

(9) $q : (B, j) \to (A, i)$ is a covering of edge-indexed graphs,

in the sense of (2.5)(4). If $b \epsilon VB$ and $q(b) = a$, we can identify X with $(\widetilde{B, j, b})$, so that we have a commutative diagram of projections,

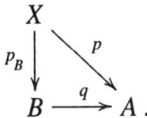

It follows that, assuming (9),

(10)   If $\mathbf{B}$ is a faithful grouping of (B, j), then $\pi_1(\mathbf{B}, b) \le G_H$.

Following are two basic theorems about G_H.

3.4   Conjugacy Theorem ([B3], Theorem (5.2)). *If $\Gamma \le G_H$ acts freely on X, then $g\Gamma g^{-1} \le H$ for some $g \epsilon G_H$.*

3.5   Discreteness Criterion; Rigidity of (A, i)

We next quote a criterion for G_H $(= G_{(A,i)})$ to be discrete, in which case we say that (A, i) is *rigid*. We call the *tree X-rigid* if G itself is discrete, i.e., if $I(G \backslash\backslash X)$ is rigid. Call (A, i) *unramified* if $i(e) = 1$ for all $e \epsilon EA$, and *discretely ramified* if it satisfies, for all $e \epsilon EA$:

(1) $i(e) > 1 \Rightarrow \begin{cases} i(e) = 2, e \text{ is a separating edge of } A, \\ \text{and } (A_0(e), i) \text{ is an unramified tree (cf. (2.1)).} \end{cases}$

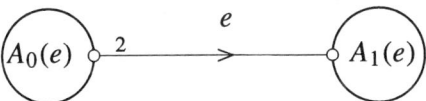

Assume that (A, i) is discretely ramified and $EA \ne \emptyset$. One possibility is

$$(A, i) = \underbrace{A_0(e)}\ \xrightarrow[\quad 2 \quad]{e}\ \underbrace{A_1(e)}$$

where $A_0(e)$ and $A_1(e)$ are both unramified trees. Apart from this case, (A, i) must have the form

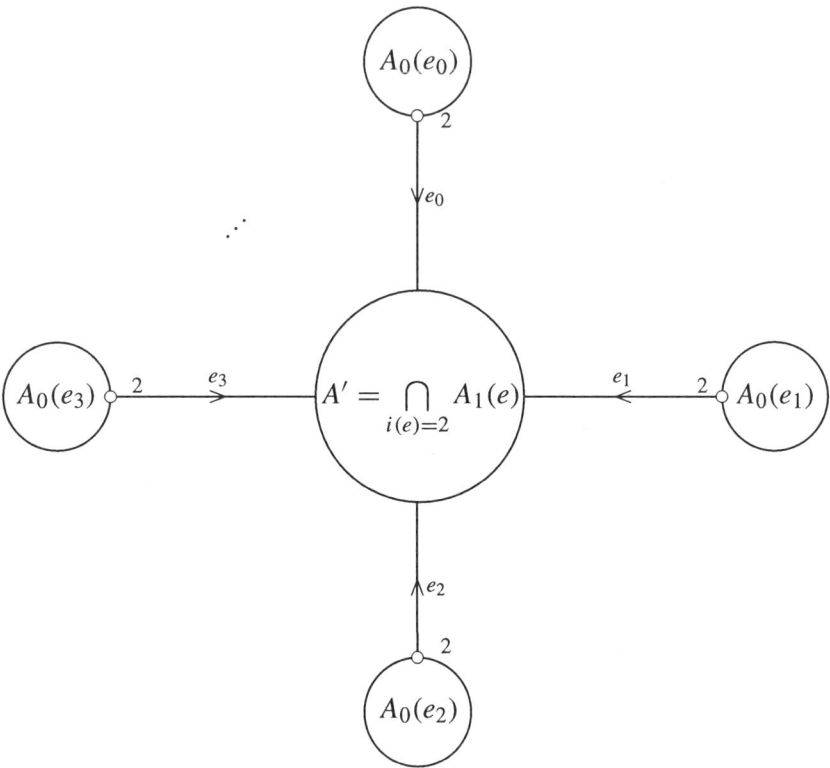

where each $A_0(e_j)$ $(j = 0, 1, 2, 3, \dots)$ is an unramified tree, and (A', i) is an un-ramified graph. It follows easily that (A, i) is unimodular, and that for $a \in VA'$ and $b \in VA$, we have

$$\frac{\Delta b}{\Delta a} = \begin{cases} 1 & \text{if } b \in VA' \\ 2 & \text{if } b \in VA_0(e_j) \text{ for some } j. \end{cases}$$

Now from (2.6)(11) it follows that

Corollary. *If (A, i) is discretely ramified, then it is unimodular. If further* Vol$((A, i))$ $< \infty$, *then A is a finite graph.*

In case H (or, equivalently, G_H) acts minimally on X, then (A, i) is "minimal" (cf. [B3], (7.12)), and then (1) above takes the simpler form:

$$(1)_{\min} \qquad\qquad i(e) > 1 \Rightarrow \begin{cases} i(e) = 2, \text{ and } \partial_0 e \text{ is a} \\ \text{terminal vertex of } A. \end{cases}$$

Theorem (Appendix [BT], Theorem (1.3)). *Let $H \leq G = \mathrm{Aut}(X)$ act without inversions on X, and put $(A, i) = I(H \backslash\backslash X) = I(G_H \backslash\backslash X)$.*

(a) *If (A, i) is discretely ramified, then $H = G_H$, H is discrete (so (A, i) is rigid), and $H \cong F * T$, where $F = \pi_1(A)$, a free group, and T is a free product of groups of order 2. If H contains an X-lattice, Γ, then $[H : \Gamma] < \infty$, and H and Γ are uniform X-lattices.*

(b) *G_H is discrete (i.e., (A, i) is rigid) iff either G_H is finite or (A, i) is discretely ramified.*

In case H acts with inversions on X, then we can pass to the tree X' obtained by subdividing each edge of X inverted by H. Then we see that G_H is discrete iff either G_H is finite or $I(H \backslash\backslash X')$ is discretely ramified (cf. Appendix [BT]).

Consider the graph $A = \circ\!\!\underset{-1}{}\!\!\overset{}{\rule{1.5cm}{0.4pt}}\!\!\underset{0}{}\!\!\overset{}{\rule{1.5cm}{0.4pt}}\!\!\underset{1}{\circ}$, with covering tree

$$X = \tilde{A} = \cdots$$ 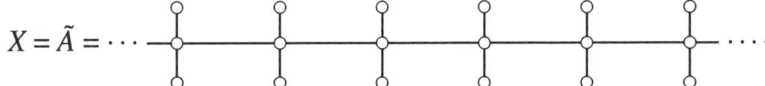 $$\cdots$$

Identifying VX naturally with $\mathbf{Z} \times \{-1, 0, 1\}$, we have $\sigma, \tau, \rho \epsilon G = \mathrm{Aut}(X)$, defined by $\sigma(n, \varepsilon) = (n, -\varepsilon)$, $\tau(n, \varepsilon) = (n + 1, \varepsilon)$, and $\rho(n, \varepsilon) = (-n, \varepsilon)$. Then $\pi_1(A) = \langle \tau \rangle$ is an X-lattice, and all X-lattices are uniform. But G is not discrete, i.e., X is not rigid. In fact, the group $H = \langle \sigma \rangle$ is finite (of order 2), yet G_H ($\cong \{\pm 1\}^{\mathbf{Z}}$) is not discrete. We have $G = G_H \rtimes D_\infty$, with $D_\infty = \langle \tau, \rho \rangle$. Two X-lattices need not be commensurable.

The relevant phenomenon here is that the central axis, X_0, of X is a G-invariant subtree that is rigid (so $G|X_0$ is discrete), with quotient $A_0 = \langle \tau \rangle \backslash X_0 = \underset{0}{\circ} \subset A$.

More generally, given any (A, i) and $X = \widetilde{(A, i)}$, we shall say that (A, i) is *virtually rigid* if there is a subtree $X_0 \subset X$ invariant under $H = G_{(A,i)}$, and such that X_0 is rigid; hence $H|X_0$ is discrete. We then have $(A_0, i) = I(H \backslash\backslash X_0) \subset (A, i)$. A criterion ("$\ell(H) \neq 0$") for the existence of a unique minimal H-invariant subtree X_0 is given in (5.7) below.

3.6 Unimodularity and volume

It is shown in [BK], §4 that $(A, i) = I(H \backslash\backslash X)$ is unimodular (in the sense of (2.6)(4)) iff the locally compact group $\bar{H}$ is unimodular. In this case, a Haar measure μ on $\bar{H}$ is invariant under conjugation. Hence we can define $N : VA \to \mathbf{R}^\times$ by $N(p(x)) = \mu(\bar{H}_x)$, for $x \epsilon VX$, and $p : X \to A = H \backslash X = \bar{H} \backslash X$. This function N satisfies the condition (2.6)(4)(b) since $N(e) = \mu(\bar{H}_e) = \mu(\bar{H}_{\bar{e}}) = N(\bar{e})$. Hence $N = \mu(\bar{H}_x) \cdot N_a$

(cf. (2.6)(10)) if $p(x) = a$. Then we have

$$\mu(\bar{H}\backslash\backslash X) := \sum_{x \in \bar{H}\backslash VX} 1/\mu(\bar{H}_x)$$

(1)

$$= (1/\mu(\bar{H}_x)) \cdot \text{Vol}_a((A, i)) \qquad \text{if } p(x) = a.$$

From (1.6) we have:

(2)   If $\Gamma \leq \bar{H}$ is a discrete subgroup, then Γ is an X-lattice iff Γ is an $\bar{H}$-lattice, and $\mu(\bar{H}\backslash\backslash X) < \infty$. In this case,

$$\text{Vol}(\Gamma\backslash\backslash X) = \mu_{\Gamma\backslash\bar{H}}(\Gamma\backslash\bar{H}) \cdot \mu(\bar{H}\backslash\backslash X).$$

Combining (3.3)(7) and (2.6)(19) we have (cf. [BK], (4.6))

3.7   Theorem. *Let $H \leq G = \text{Aut}(X)$ be a subgroup without inversions, and $(A, i) = I(H\backslash\backslash X)$. The following conditions are equivalent:*

(a) There is a discrete subgroup $\Phi \leq G_H$ such that $\Phi\backslash X = H\backslash X \;(= A)$.

(b) (A, i) has bounded denominators, in the sense of (2.6)(18).

Under these conditions, Φ is a uniform G_H-lattice. Moreover Φ is an X-lattice iff $\text{Vol}((A, i)) < \infty$, and Φ is a uniform X-lattice iff A is finite.

3.8   Existence of tree lattices

Consider the conditions

   (U) (A, i) is unimodular;   (BD) (A, i) has bounded denominators;

   (FV) $\text{Vol}(A, i) < \infty$;   (F) A is finite.

Recall from (2.6)(18) that (BD) $\Rightarrow$ (U), and conversely in the presence of (F). Theorem (3.7) asserts that

$$[(U) + (BD)] \Longleftrightarrow [\exists \text{ a discrete } \Phi \leq G_{(A,i)} \text{ such that } \Phi\backslash X = A],$$

and such a Φ must be a uniform $G_{(A,i)}$-lattice. Further,

$$[\Phi \text{ is an } X\text{-lattice}] \Longleftrightarrow (\text{FV}),$$

$$[\Phi \text{ is a uniform } X\text{-lattice}] \Longleftrightarrow (\text{F}).$$

3.9   Lattice Existence Theorem (Appendix [BCR]).

$$[\exists \text{ an } X\text{-lattice } \Gamma \leq G_{(A,i)}] \Longleftrightarrow [(U) + (FV)].$$

Moreover, Γ may be chosen to be a uniform $G_{(A,i)}$-lattice.

There remains the question of whether one can also choose Γ above to be a non-uniform $G_{(A,i)}$-lattice. This is impossible if (A,i) is rigid (i.e., $G_{(A,i)}$ is discrete), and, more generally, if (A,i) is virtually rigid, in the sense of (3.5). For then all X-lattices in $G_{(A,i)}$ are uniform X-lattices. On the other hand, there is some evidence that this is the only obstacle. In [CR1], Carbone and Rosenberg investigate whether, when $G_{(A,i)}$ admits a non-uniform X-lattice, it also admits a non-uniform $G_{(A,i)}$-lattice.

The uniform case is covered by:

3.10   **Theorem** ([BK], §4). *With notation as in (3.7), the following conditions are equivalent:*

   (a) (U) (A,i) *is unimodular, and* (F) A *is finite.*
   (b) *There is a uniform X-lattice* $\Phi \leq G_H$ *such that* $\Phi \backslash X = A$.
   (c) *There is a free uniform X-lattice* $\Gamma \leq H$.

Applied to $H = G$ (or rather to an index 2 subgroup without inversions), cf. (4.1)(6) below, this gives:

3.11   **Corollary** ([BK], (4.10)). *The following conditions are equivalent:*

   (a) G *is unimodular and* $G \backslash X$ *is finite.*
   (b) *There exists a uniform X-lattice.*
   (c) X *is the universal cover of a finite connected graph.*

Under the conditions of (3.11) X is called a *uniform tree.*

3.12   The structure of tree lattices

We now survey some of the results on tree lattices proved here and elsewhere. Let X be a locally finite tree, and $\Gamma \leq G = \mathrm{Aut}(X)$ an X-lattice ((3.1)), i.e., Γ is discrete and

$$(*) \qquad \mathrm{Vol}(\Gamma \backslash\backslash X) \quad \left(= \sum_{x \in \Gamma \backslash VX} 1/|\Gamma_x| \right) \quad < \infty.$$

One of three cases occurs (cf. (5.1), (5.7), (9.7)).

(F) X is finite: Then Γ is finite, and uniform. Γ fixes a vertex or inverts a (unique) edge.

(P) X is "parabolic," i.e., has a unique end ε: Then Γ is infinite and locally finite, and non-uniform. Each horoball centered at ε is Γ-invariant.

$(\ell(\Gamma) \neq 0)$ X is infinite and non-parabolic: Then Γ is infinite, and there is a unique minimum Γ-invariant subtree X_Γ.

Uniform tree lattices. Suppose that Γ is uniform

- $\Gamma \backslash X$ is finite.
- Γ is finitely generated ((5.6) below) and virtually free (cf. [BK], (2.8)), hence residually finite.
- $Z_G(\Gamma)$ is finite, unless Γ is virtually infinite cyclic (cf. (6.6)).
- $N_G(\Gamma)/\Gamma$ is finite (cf. [BK], (6.4)).
- $C_G(\Gamma)$, the commensurator of Γ, is dense in G ([Li1]).
- $\mathrm{Vol}(\Gamma \backslash\backslash X)$ is a rational number whose denominator involves only primes $\leq d = \mathrm{Max}_{x \in VX} \deg(x)$, and not divisible by d^2 if d is prime. For $X = X_d$, every such rational number > 0 occurs as $\mathrm{Vol}(\Gamma \backslash\backslash X)$ for some Γ ([Le]; see also [R] and (0.6)(3) for the biregular case).
- Let $X = X_d$, the regular tree of degree d. For $d \geq 3$, there exist infinite ascending chains, $\Gamma_1 < \Gamma_2 < \Gamma_3 < \cdots$, of uniform X-lattices; hence $\lim_{n \to \infty} \mathrm{Vol}(\Gamma_n \backslash\backslash X) = 0$. (cf. (0.7)(4).) For $d \geq 5$, there exist $v > 0$ for which there are infinitely many conjugacy classes of uniform X- lattices Γ with $\mathrm{Vol}(\Gamma \backslash\backslash X) = v$ (cf. [BK], (7.1)).
- There exists a uniform tree X for which $G = \mathrm{Aut}(X)$ is itself a non-abelian free uniform X-lattice, i.e., G acts freely without inversion on X and $G \backslash X$ is finite (cf. [BK], (4.12)(2), and Appendix [BT]).
- If X is infinite then $\ell(\Gamma) \neq 0$ and $X_\Gamma = X_G$. ((5.12)).

Non-uniform tree lattices. Suppose that Γ is non-uniform.

- $\Gamma \backslash X$ is infinite. The number of cusps may be finite or infinite ((4.13) and (4.16)). The rank of the free group $\pi_1(\Gamma \backslash X)$ can be any cardinal $\leq \aleph_0$ ((4.2)(7) and (4.17)).
- Γ is not finitely generated ((5.16)). It has finite subgroups of unbounded orders. Hence Γ is not virtually torsion free. Γ may or may not be residually finite ((10.2) and (10.3)).
- If Γ is not locally finite or virtually cyclic then $Z_G(\Gamma)$ is a product (finite or infinite) of finite groups ((6.7)).
- $N_G(\Gamma)/\Gamma$ is a profinite group. It may be infinite or finite (even trivial) even when $X = X_d$ ($d \geq 3$).
- $C_G(\Gamma)$ is may or may not be dense in G; on X_d ($d \geq 3$) both cases occur.
- If X is uniform and admits a non-uniform lattice, then any real number $v > 0$ is of the form $v = \mathrm{Vol}(\Gamma \backslash\backslash X)$ for some non-uniform X-lattice Γ. This is proved here ((4.3)) for X_d ($d \geq 3$), and in general in [R].
- G acts minimally on X iff X has no terminal vertices. In this case Γ also acts minimally on X, and $\ell(\Gamma) \neq 0$ ((5.12)).
- If $\ell(\Gamma) = 0$, then $\ell(G) = 0$ and X is a parabolic tree ((9.7)).

3.13   **Corollary.** *Let H be a simple rank 1 Lie group over a non-archimedean locally compact field F, with Bruhat–Tits tree X, and $G = \mathrm{Aut}(X)$. Let $\Gamma \leq H < G$ be a non-uniform H-lattice. Then Γ has unbounded torsion, so char $(F) > 0$, and Γ is not finitely generated. Moreover $N_H(\Gamma)/\Gamma$ is finite, while the profinite group $N_G(\Gamma)/\Gamma$ may be infinite $((10.2)(11)$ and $(17))$. The commensurator $C_H(\Gamma)$ is dense in H iff Γ is arithmetic ([Mar]).*

Proof. $H \leq G = \mathrm{Aut}(X)$ is a closed unimodular subgroup without inversions, $H \backslash X$ is finite, and so Γ is a non-uniform X-lattice. Hence Γ has unbounded torsion (to make $\mathrm{Vol}(\Gamma \backslash\backslash X) < \infty$) and Γ is not finitely generated by (5.4)(2).

3.14   Non-arithmetic uniform commensurators

Let H be a simple rank 1 Lie group over a non-archimedean local field F, with Bruhat–Tits tree X, and $G = \mathrm{Aut}(X)$. Let $\Gamma \leq H < G$ be a uniform H-lattice. Then $C_G(\Gamma)$ is dense in G ([Li1]). On the other hand, if Γ is non-arithmetic (such examples exist, by [L3]) then, by [Mar], $C_H(\Gamma)$ is discrete! Thus $C_H(\Gamma) = H \cap C_G(\Gamma)$ is a lattice in H, while $C_G(\Gamma)$ is dense in G.

Chapter 4

Arbitrary Real Volumes, Cusps, and Homology

4.0 Introduction

Let X be a locally finite tree and Γ a non-uniform X-lattice. In this section we show that, even when X is regular (of degree ≥ 3), $\mathrm{Vol}(\Gamma\backslash\backslash X)$ can take any positive real value ((4.3)), that $\Gamma\backslash X$ can have any conceivable number of "cusps" ((4.13) and (4.16)), and that these phenomena can occur with $\pi_1(\Gamma\backslash X)$ either finitely or infinitely generated. If we drop regularity, then every locally finite connected graph can occur as some $\Gamma\backslash X$ ((4.17)).

From Chapter 2 we know that Γ and X are determined by the faithful graph of finite groups,

$$\begin{aligned} \mathbf{A} &= (A, \mathcal{A}) = \Gamma\backslash\backslash X, \\ p &: X \to A = \Gamma\backslash X. \end{aligned}$$

(1)

Put

$$(A, i) = I(\mathbf{A}).$$

(2)

For $a_0 \in VA$ we then have

$$\Gamma = \pi_1(\mathbf{A}, a_0) \quad \text{and} \quad X = \widetilde{(A, i, a_0)}.$$

(3)

For $x \in VX$, $p(x) = a \in VA$, we have

$$\deg_X(x) = \deg_{(A,i)}(a) := \sum_{e \in E_0(a)} i(e).$$

(4)

Thus

(5)
$$X = X_d, \ \text{the } d\text{-regular tree, iff}$$

$$\deg_{(A,i)}(a) = d \ \ \forall \, a \epsilon VA, \ \ \text{i.e., ``}(A, i) \text{ is } d\text{-regular.''}$$

Our constructions will generally start by constructing (A, i), and then use $(2.6)(19)$ and (15) to obtain $\mathbf{A}$, and hence the X-lattice Γ. We now briefly review these criteria.

For $e \epsilon EA$ put $\triangle(e) = i(\bar{e})/i(e)$. For an edge path $\gamma = (e_1, \ldots, e_n)$, put $\triangle(\gamma) = \triangle(e_1) \ldots \triangle(e_n)$. We call (A, i) *unimodular* if $\triangle(\gamma)$ depends only on the end points $a = \partial_0 e_1$ and $b = \partial_1 e_n$, in which case $\triangle\gamma$ is denoted $\triangle b/\triangle a$. Fix a base point $a_0 \epsilon VA$ and put $N = N_{a_0}$: $N(a) = \triangle a/\triangle a_0$ for $a \epsilon VA$. For $e \epsilon E_0(a)$ we also put $N(e) = N(a)/i(e)$. We say that (A, i) has *"bounded denominators"* (BD) if $((A, i)$ is unimodular and$)$, the rational numbers $N(e)(e \epsilon EA)$ have bounded denominators. Then $(2.6)(19)$ says that

(6) (A, i) admits a (faithful) finite grouping $\mathbf{A}$ iff
       (BD) (A, i) has bounded denominators.

We further have

(7)
$$\mathrm{Vol}_{a_0}(A, i) = \sum_{a \epsilon VA} \frac{1}{N(a)},$$

and $\Gamma = \pi_1(\mathbf{A}, a_0)$ will be an X-lattice iff this $\mathrm{Vol}_{a_0}(A, i)$ is finite. In fact $((2.6)(15))$

(8)
$$\mathrm{Vol}(\Gamma \backslash\backslash X) = \mathrm{Vol}(\mathbf{A}) = \frac{1}{|\mathcal{A}_{a_0}|} \mathrm{Vol}_{a_0}(A, i).$$

4.1 Grafting

As in (4.0), let (A, i) be a rooted indexed graph with a faithful finite grouping $\mathbf{A}$.

(1)   For $a \epsilon VA$ put $d(a) = \deg_{(A,i)}(a) := \sum_{e \epsilon E_0(a)} i(e)$.

We want to observe the effects of grafting a rooted indexed graph (B, j) to (A, i) at a, to obtain an enlarged graph $(A^+, i) = (A, i) \cup_a (B, j)$.

(2)   Let $\overset{\textstyle B_m}{\underset{a}{\bigcirc}}$ denote a bouquet of m unramified loops $(j(f) = 1 \ \forall \ f \epsilon E B_m)$. Put $(A^+, i) = (A, i) \bigcup_a \overset{\textstyle B_m}{\underset{a}{\bigcirc}}$. Extend $\mathbf{A}$ to a grouping $\mathbf{A}^+$ of (A^+, i) with the same vertex groups. Then

$$\mathrm{Vol}(\mathbf{A}^+) = \mathrm{Vol}(\mathbf{A}),$$

$$\deg_{(A^+,i)}(a) = d(a) + 2m, \ \ \text{and}$$

$$\pi_1(\mathbf{A}^+, a_0) \cong \pi_1(\mathbf{A}, a_0) * F_m,$$

where F_m is a free group on m generators.

We can use this to expand examples to regular trees without changing volume, while making $\pi_1(A)$ become infinitely generated, as follows.

(3)   Assume that $d(a)$ $(a \epsilon VA)$ is bounded, and constant mod 2. Choose $m(a)$ $(a \epsilon VA)$ so that

$$D := d(a) + 2m(a)$$

is independent of a, and $m(a)$ is > 0 infinitely often. Let (A^+, i) be obtained from (A, i) by grafting, at each $a \epsilon VA$, a bouquet of $m(a)$ unramified loops. Extend $\mathbf{A}$ to a grouping $\mathbf{A}^+$ of (A^+, i) with the same vertex groups. Then

$$\mathrm{Vol}(\mathbf{A}^+) = \mathrm{Vol}(\mathbf{A}),$$

$$(\widetilde{A^+, i}, a_0) = X_D \quad \text{(the D-regular tree),} \quad and$$

$$\pi_1(\mathbf{A}^+) \text{ is not finitely generated.}$$

Another possibility is to graft to (A, i) at $a \epsilon VA$, a "bush,"

(4) $(B, j) = B(r, D) :=$

$$j(e_h) = 1, \qquad j(\bar{e}_h) = D \quad (h = 1, \dots, r),$$
$$(A^+, i) = (A, i) \bigcup_a (B, j).$$

Choosing a group $\mathcal{A}_{a^+}$ of order D, we can then extend $\mathbf{A}$ on (A, i) to $\mathbf{A}^+ = (A^+, \mathcal{A}^+)$ by keeping the same groups in A, and putting $\mathcal{A}_{a_h^+} = \mathcal{A}_{a^+} \times \mathcal{A}_a$, and $\mathcal{A}_{e_h} = \mathcal{A}_a$ $(h = 1, \dots, r)$. This done, we have

$$\mathrm{Vol}(\mathbf{A}^+) = \mathrm{Vol}(\mathbf{A}) + \frac{r}{D|\mathcal{A}_a|},$$

(5) $$\deg_{(A^+, i)}(a) = d(a) + r, \quad \deg_{(A^+, i)}(a_h^+) = D, \quad and$$

$$\pi_1(\mathbf{A}^+, a_0) = \pi_1(\mathbf{A}, a_0).$$

Of course, $\pi_1(\mathbf{A}^+)$ will be larger than $\pi_1(\mathbf{A})$. We can use this to expand examples to regular trees, without changing the topological $\pi_1(A)$, as follows.

(6)   Suppose that $d(a)$ $(a \epsilon VA)$ is bounded by D. Let (A^+, i) be obtained from (A, i) by grafting, at each $a \epsilon VA$, a copy of $B(r(a), D)$, $r(a) = D - d(a)$

(and $B(r(a), D) = \{a\}$ if $r(a) = 0$). Extend $\mathbf{A}$ on (A, i) to $\mathbf{A}^+$ on (A^+, i), as above. Then

$$\pi_1(\mathbf{A}^+, a_0) = \pi_1(\mathbf{A}, a_0),$$

$$(\widetilde{A^+, i}, a_0) = X_D \quad \text{(the } D\text{-regular tree),} \quad \text{and}$$

$$\mathrm{Vol}(\mathbf{A}^+) = \mathrm{Vol}(\mathbf{A}) + \sum_{a \in VA} \frac{r(a)}{D|A_a|} < 2\mathrm{Vol}(\mathbf{A}).$$

Finally, we indicate how to construct a d-regular $\mathbf{A}$ $(d \geq 3))$ with $\pi_1(A)$ of any prescribed rank.

(7)   Given $d \geq 3$, and $0 \leq r \leq \aleph_0$, there exists a d-regular graph of finite groups
       $\mathbf{A}$ such that $\pi_1(A)$ is free on r generators.

Writing $d = q + 1, q \geq 2$, let us abbreviate,

$$
\begin{array}{ccccccc}
& L & & 1 & q-1 & 1 & & & 1 & q-1 & q \\
\circ & \!\!\!\!\!\!\!\!\!\!\!\!\text{---} & \circ & = & \circ\!\!\!\!\text{---}\!\!\!\!\circ & & \circ & & \circ\!\!\!\!\text{---}\!\!\!\!\circ \\
& & & & 1 & & 1
\end{array}
$$

If $r = \aleph_0$, put

$$
(A, i) = \quad \overset{d}{\circ}\!\!\!\!\!\!\!\!\!\underset{}{\text{---}}\!\!\!\!\!\!\overset{q}{\circ}\!\!\!\!\!\!\!\!\!\!\!\!\overset{L}{\text{---}}\!\!\!\!\!\!\circ \;\cdots\; \circ\!\!\!\!\!\!\!\!\overset{L}{\text{---}}\!\!\!\!\!\!\circ\!\!\!\!\!\!\!\!\overset{L}{\text{---}}\!\!\!\!\!\!\circ \;\cdots.
$$

If $r < \aleph_0$, put

$$
(A, i) = \quad \circ\!\!\!\!\!\!\!\!\overset{d}{\text{---}}\!\!\!\!\!\!\overset{q}{\circ}\!\!\!\!\!\!\!\!\overset{L}{\text{---}}\!\!\!\!\!\!\circ \;\cdots\; \circ\!\!\!\!\!\!\!\!\overset{L}{\text{---}}\!\!\!\!\!\!\circ\!\!\!\!\!\!\overset{1}{}\!\!\!\!\!\!\overset{q}{}\!\!\!\!\!\!\circ\!\!\!\!\!\!\overset{1}{}\!\!\!\!\!\!\overset{q}{}\!\!\!\!\!\!\circ \;\cdots.
$$
$$\text{(}r\text{ copies of } L\text{)}$$

Clearly (A, i) is d-regular and unimodular and $\pi_1(A)$ is free on r generators. It is also easily seen that (A, i) has bounded denominators, and finite volume. Hence (A, i) admits a finite grouping $\mathbf{A}$, by (2.6)(19), whence (7).

4.2   Volumes

Let X be a uniform tree, $G = \mathrm{Aut}(X)$ and

(1) $$d = \sup_{x \in VX} \deg_X(x) \quad (< \infty).$$

For $x \in VX$, G_x is an iterated wreath product of symmetric groups S_q, with $q \leq d$, and S_d can occur only if $\deg(x) = d$, in which case it occurs only once (cf. (6.16)). It follows that the order of a finite subgroup of G_x is divisible only by primes $\leq d$, and, when d itself is prime, the order is not divisible by d^2. Since any finite subgroup of G fixes a vertex of (the barycentric subdivision of) X it follows that:

(2)   If $\Gamma \leq G$ is a uniform X-lattice, then $\mathrm{Vol}(\Gamma \backslash\!\backslash X) = a/b$ $(a, b \in \mathbf{N})$, where b is a "d-number," by which we mean that b is divisible only by primes $p \leq d$ and, if d is prime, $d^2 \nmid b$.

Gabe Rosenberg [R] has shown:

(3)   Let $X = X_{d,d'}$ be the (d, d')-biregular tree, of degrees $d \geq d' \geq 2$. Let $v = N/D > 0$ be a rational number in reduced form. Then $v = \mathrm{Vol}(\Gamma \backslash\!\backslash X)$ for some uniform X-lattice Γ iff D is a d-number, and, if $d + d'$ is prime, then $(d + d')|N$.

When $d = d'$, so $X = X_d$, this result was proved by Inga Levich [Le]. We show now that, in contrast, for $d \geq 3$, $\mathrm{Vol}(\Gamma \backslash\!\backslash X_d)$ can take any real value > 0 for non-uniform lattices Γ.

4.3 **Theorem.** *Given an integer $d \geq 3$ and any real number $v > 0$, there exists a lattice Γ on the d regular tree $X = X_d$ such that* $\mathrm{Vol}(\Gamma \backslash\!\backslash X) = v$, *and*

$$\Gamma \backslash X = \circ\!\!-\!\!-\!\!-\!\!-\!\!\circ\!\!-\!\!-\!\!-\!\!-\!\!\circ\!\!-\!\!-\!\!-\!\!-\!\!\circ\!\!-\!\!-\!\!-\!\!-\!\!\circ \cdots$$

Theorem (4.3), without the claim about $\Gamma \backslash X$, has been extended by Gabe Rosenberg ([R]) to any uniform tree that admits a non-uniform lattice (cf. (0.5)(13)).

4.4 **Remark.** Let $(A, i) = I(\Gamma \backslash\!\backslash X)$. Let (A^+, i) be the result of grafting to each $a \in VA$ a bouquet of m unramified loops, as in (4.1)(3). Then

$$X^+ := \widehat{(A^+, i)} = X_{d+2m},$$

and $\mathbf{A} = \Gamma \backslash\!\backslash X$ extends to a grouping $\mathbf{A}^+$ of (A^+, i) which defines an X^+-lattice Γ^+ such that $\mathrm{Vol}(\Gamma^+ \backslash\!\backslash X^+) = \mathrm{Vol}(\Gamma \backslash\!\backslash X) = v$ and $\pi_1(\Gamma^+ \backslash X^+) = \pi_1(A^+)$ is an infinitely generated free group.

4.5 **Proof of (4.3).** We start by constructing a d-regular edge-indexed graph (A, i) based on the graph

(1) $A = $

$$\begin{array}{ccccccccccc}
& e_0 & & e_1 & & e_2 & & e_3 & & e_4 & \\
\circ\!\!\!\!&\longrightarrow&\!\!\!\!\circ\!\!\!\!&\longrightarrow&\!\!\!\!\circ\!\!\!\!&\longrightarrow&\!\!\!\!\circ\!\!\!\!&\longrightarrow&\!\!\!\!\circ\!\!\!\!&\longrightarrow&\!\!\!\!\circ \cdots \\
a_{-1} & & a_0 & & a_1 & & a_2 & & a_3 & & a_4
\end{array}$$

The indexing will satisfy the following conditions. Write

$$d = p + 1, \qquad (p \geq 2).$$

(2) $$(i(e_0), i(\bar{e}_0)) = (p + 1, p).$$

(3) $$\text{For all } n \geq 0, \quad \deg_{(A,i)}(a_n) \; (= i(\bar{e}_n) + i(e_{n+1})) = p + 1.$$

(4)   For all $n > 0$,

$$(i(e_n), i(\bar{e}_n)) = \begin{cases} (1, 1) & \text{so } \Delta(e_n) = 1, \text{ or} \\ (p, p) & \text{so } \Delta(e_n) = 1, \text{ or} \\ (1, p) & \text{so } \Delta(e_n) = p. \end{cases}$$

(5)   The indices n for which $(i(e_n), i(\bar{e}_n)) = (1, p)$ form an infinite sequence

$$1 \leq n_1 < n_2 < n_3 < \cdots .$$

(We agree to put $n_0 = 0$.)

Note that conditions (2) and (3) imply that $i(e_1) = 1$, and that (A, i) is d-regular. Further, the indexed segment from e_{n_r} to $e_{n_{r+1}}$ must have the form

It follows that,

(6) $$c_r := n_{r+1} - n_r \text{ is odd, for all } r \geq 0.$$

Further, we can choose the sequence (n_r) in (5) arbitrarily, provided only that the c_r in (6) are all odd.

We have

(7) $$\Delta(e_0) = \frac{p}{p+1}, \quad \Delta(e_{n_r}) = p, \quad \text{and} \quad \Delta(e_n) = 1 \text{ otherwise.}$$

Thus, for $n \geq 0$, we have

$$\frac{\Delta a_n}{\Delta a_{-1}} = \Delta(e_0)\Delta(e_1)\ldots, \Delta(e_n)$$

$$= \frac{p}{p+1}\Delta(e_1)\ldots\Delta(e_n);$$

so, in view of (7)

(8) $$\frac{\Delta a_n}{\Delta a_{-1}} = \frac{p^{r+1}}{p+1} \quad \text{for } n_r \leq n < n_{r+1}.$$

Put $V = \mathrm{Vol}_{a_{-1}}(A, i)$. (See (2.6)(11).) Thus

$$V = \sum_{n \geq -1} \frac{\Delta a_{-1}}{\Delta a_n} = 1 + \sum_{n \geq 0} \frac{\Delta a_{-1}}{\Delta a_n} = 1 + \sum_{r \geq 0} c_r \frac{\Delta a_{-1}}{\Delta a_{n_r}}, \quad so$$

(9) $$V = 1 + (p+1) \sum_{r \geq 0} \frac{c_r}{p^{r+1}}.$$

Next we construct a grouping $\mathbf{A} = (A, \mathcal{A})$ of (A, i) as follows. Let H be a group of order $p + 1$, and let

(10) $$Q_0 < Q_1 < Q_2 < \cdots$$

be a chain of groups with $|Q_m| = p^m$, and put $Q = \bigcup_m Q_m$. Fix some $m_0 > 0$. Put

(11) $$\mathcal{A}_{a_{-1}} = H \times Q_{m_0 - 1},$$

and, for $n \geq 0$, say $n_r \leq n < n_{r+1}$,

(12) $$\mathcal{A}_{a_n} = Q_{m_n}, \quad \text{where } m_n = m_0 + r,$$
$$\mathcal{A}_{e_n} = \begin{cases} Q_{m_n} & \text{if } i(\bar{e}_n) = 1 \\ Q_{m_n - 1} & \text{if } i(\bar{e}_n) = p \end{cases}.$$

The maps of edge groups into adjacent vertex groups are inclusions. It is easily seen that $\mathcal{A}$ gives a grouping of (A, i), and so, for $n \geq 0$,

$$|\mathcal{A}_{a_n}| = |\mathcal{A}_{a_{-1}}| \cdot \frac{\Delta a_n}{\Delta a_{-1}}$$

$$= (p+1)p^{m_0 - 1} \cdot \frac{p^{r+1}}{p+1} \quad \text{for } n_r \leq n < n_{r+1}$$

$$= p^{m_0 + r} = p^{m_n}.$$

It follows that

(13) $$\mathrm{Vol}(\mathbf{A}) = \frac{1}{|\mathcal{A}_{a_{-1}}|} \cdot V \quad \text{(cf. (9))}$$
$$= \frac{1}{(p+1)p^{m_0 - 1}} + \sum_{r \geq 0} \frac{c_r}{p^{m_0 + r}}.$$

The proof of (4.3) will be completed once we show the following two things:

4.6 Lemma. *Given a real number $v > 0$, and an integer $p \geq 2$, we can choose integers $m_0 > 0$ and odd integers $c_r > 0$ $(r \geq 0)$ so that*

$$v = \frac{1}{(p+1)p^{m_0 - 1}} + \sum_{r \geq 0} \frac{c_r}{p^{m_0 + r}}.$$

4.7 **Lemma.** *We can choose the groups Q_m in (4.5)(10) so that $Q = \bigcup_m Q_m$ contains no finite normal subgroups $\neq \{1\}$.*

In fact (4.5)(13) and (4.6) assure us that $\mathrm{Vol}(\mathbf{A}) = v$, while Lemma (4.7) makes $\mathbf{A}$ faithful (cf. [B3], (1.23)), and so $\Gamma = \pi_1(\mathbf{A}, a_{-1})$ is a subgroup of $G = \mathrm{Aut}(X)$, $X = (\widetilde{\mathbf{A}, i, a_{-1}}) = X_d$, (by (4.5)(3)) with $\Gamma \backslash\backslash X = \mathbf{A}$, and thus Γ furnishes the desired lattice with $\mathrm{Vol}(\Gamma \backslash\backslash X_d) = v$.

Proof of (4.6). Choose $m_0 > 0$ large enough so that

$$s := p^{m_0} v - \frac{p}{p+1} \geq 3.$$

Then we seek odd integers $c_r > 0$ $(r \geq 0)$ so that

$$s = \sum_{r \geq 0} \frac{c_r}{p^r} \quad \left(\text{hence } v = \frac{p}{p^{m_0}(p+1)} + \sum_{r \geq 0} \frac{c_r}{p^{m_0+r}} \right).$$

For this first write $s = \sum_{r \geq 0} \frac{a_r}{p^r}$, where the a_r are integers with $0 \leq a_r < p$ for $r > 0$, and $a_0 \geq 2$ since $s \geq 3$. Since

$$(p-1) \left(\sum_{r>0} \frac{1}{p^r} \right) = 1$$

we have

$$s = a_0 + \sum_{r>0} \frac{a_r}{p^r}$$

$$= (a_0 - 1) + \sum_{r>0} \frac{a_r + (p-1)}{p^r}$$

$$= d_0 + \sum_{r>0} \frac{d_r}{p^r}, \quad \text{with } d_r > 0 \; \forall \, r \geq 0.$$

If all d_r are odd, we are done. If not, let r_1 be the least r for which d_r is even. Thus $d_{r_1} \geq 2$. Now replace $\frac{d_{r_1}}{p^{r_1}}$ by

$$\frac{d_{r_1} - 1}{p^{r_1}} + \frac{1}{p^{r_1}} = \frac{d_{r_1} - 1}{p^{r_1}} + \sum_{r>r_1} \frac{p-1}{p^r}.$$

With

$$d'_r = \begin{cases} c_r := d_r & \text{for } r < r_1 \\ c_{r_1} := d_{r_1} - 1 & \text{for } r = r_1 \\ d_r + (p-1) & \text{for } r > r_1 \end{cases}$$

we have $d'_r > 0$ for all r, $d'_r = c_r$ is odd for $r \leq r_1$ and $s = \sum_{r \geq 0} \frac{d'_r}{p^r}$. Now we continue indefinitely in this fashion, to eventually make all of the coefficients odd, and so obtaining the desired expansion $s = \sum_{r \geq 0} \frac{c_r}{p^r}$.

Proof of (4.7). Let R be a finite product of finite fields so that $|R| = p$. (Just write $p = p_1^{m_1}, \ldots, p_r^{m_r}$, with p_i prime and $m_i > 0$, and take $R = \mathbf{F}_{p_1^{m_1}} \times \cdots \times \mathbf{F}_{p_r^{m_r}}$.) Let

$$Q = U_\infty(R) = \begin{bmatrix} 1 & & & \\ & 1 & * & \\ & & 1 & \\ 0 & & & \ddots \end{bmatrix} \leq GL_\infty(R).$$

Note that $U_\infty(R)$ is the ascending union of the groups

$$U_n(R) = \begin{bmatrix} 1 & & & \\ & 1 & * & \\ & & \ddots & \\ 0 & & & 1 \end{bmatrix} \leq GL_n(R).$$

Each $U_n(R)$ has a chain of normal subgroups with successive quotients isomorphic to R, of order p, and which refines the chain

$$1 < U_2(R) < U_3(R) < \cdots < U_n(R).$$

(From $U_n(R)$ to $U_{n+1}(R)$, allow successively more entries in the last column, from top to bottom.) We can use these to define

$$1 = Q_0 < Q_1 < Q_2 < \cdots$$

with $Q_m/Q_{m-1} \cong R$ and $\bigcup_m Q_m = Q$. Now it is readily seen that Q contains no finite normal subgroup $\neq \{1\}$.

4.8 Cusps

Let X be a locally finite tree, and Γ an X-lattice, with quotient

(1)
$$p : X \to A = \Gamma \backslash X,$$
$$\mathbf{A} = (A, \mathcal{A}) = \Gamma \backslash\backslash X, \quad \text{and} \quad (A, i) = I(\mathbf{A}).$$

When Γ is non-uniform, A is infinite, and the cusps of Γ should refer, roughly speaking to the "ways of going to infinity" in A (or in $\mathbf{A}$).

Consider the classical case when Γ is a lattice in a simple rank 1 Lie group H over a non-archimedean local field (of characteristic > 0 since Γ is non-uniform (cf. (0.5)(1))) and X is the Bruhat–Tits tree of H. Then (cf. [L3]) A has the form

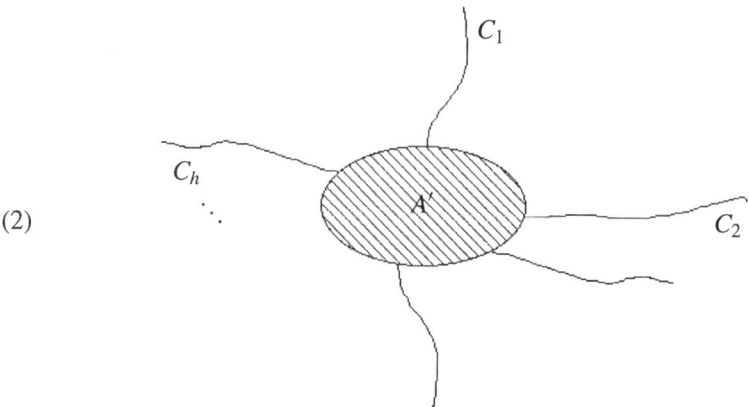

(2)

where A' is a finite graph to which is attached a finite set $\{C_1, \ldots, C_h\}$ of "isolated rays," representing the cusps. Moreover the groups $\mathcal{A}_a$ increase indefinitely as the vertex a traverses a ray C_i toward its infinite end.

In the case of a general non-uniform tree lattice Γ, as in (1) above, there are several possible generalizations of the notion of a cusp. We shall investigate two of them. One, called a *geometric parabolic end,* depends only on the geometry of A. There are at most countably many of these, they form a discrete subspace of the (profinite) space, Ends(A), of ends of A, and, if $\pi_1(A)$ is finitely generated and Ends(A) is countable, then the geometric parabolic ends are dense in Ends(A). We show that, subject to these constraints only, all possibilities are realizable with $A = \Gamma \backslash X$, Γ a non-uniform X-lattice, and $X = X_d$, the d-regular tree, for any $d \geq 3$. (Theorem (4.13)).

The other generalization of cusps is group theoretic: An end ε of X is called a Γ-*parabolic end* if the stabilizer Γ_ε fixes no end other than ε and fixes no vertex of X. It is called Γ-*cuspidal* if, further, for any ray toward ε, with vertex sequence $x_0, x_1, x_2, x_3, \ldots$, we have $\Gamma_{x_n} \leq \Gamma_\varepsilon$ for all $n \gg 0$. The Γ-cusps on X are then the Γ-cuspidal ends modulo the action of Γ. We show by examples that the number of Γ-cusps can be any cardinal from 0 to the continuum.

4.9   Geometric parabolic ends

The space $Ends(A)$ of a locally compact space A is the inverse limit of $\pi_0(A - A')$, where A' expands through compact subspaces of A. This defines a functor $A \mapsto Ends(A)$ for continuous maps $f : A \to B$ which are proper (the inverse image of a compact set is compact).

Let A be a connected locally finite graph. It is locally compact using the metric defined by minimal edge-path distance, edges having length 1. Then

(1) $$Ends(A) = \varprojlim_{A'} \pi_0(A - A'),$$

where A' expands through finite subgraphs of A. Each of the $\pi_0(A - A')$ is finite, so

(2) $Ends(A)$ is a profinite (= compact, totally disconnected) space, on which $Aut(A)$ naturally acts.

Moreover

(3) $$Ends(A) = \emptyset \quad \text{iff } A \text{ is finite.}$$

The simplest infinite example is a *ray*, i.e., a graph of the form

(4)

with unique end, denoted $end(C)$.

Two rays C, C' in A are said to have the *same direction* if the vertices of each lie within bounded distance of the other ray. This is an equivalence relation. We write $dir(C)$ for the class of C, and $Dirs(A) = \{dir(C) | C \text{ a ray in } A\}$. It is easily seen that $end(C) \in Ends(A)$ depends only on $dir(C)$, and the resulting, map

(5) $Dirs(A) \to Ends(A),$ $\qquad dir(C) \mapsto end(C)$ is surjective.

Moreover,

(6)   If $\pi_1(A)$ is finitely generated, then (5) is bijective.

To see this we first observe that:

(7) $\pi_1(A)$ is finitely generated iff A is obtained from a finite graph A_0 by attaching a rooted tree $(T(a), a)$, to each vertex $a \in VA_0$. In this case the inclusion induced map

$$[\coprod_{a \in VA_0} Ends(T(a))] \to Ends(A)$$

is bijective.

The "if" assertion follows since $\pi_1(A) = \pi_1(A_0)$. Conversely, if $\pi_1(A)$ is finitely generated it is supported on some finite $A_0 \subset A$, whence A has the indicated form. The assertion about ends is evident.

In view of (7), it suffices to prove (6) when A is a tree. In that case, if C and C' are rays in A with the same end then C and C' share all but finitely many edges, and so $dir(C) = dir(C')$, whence the injectivity of (5).

When $\pi_1(A)$ is not finitely generated (5) can fail dramatically to be injective; see Example (4.12).

A subgraph $P \subset A$ will be said to be *isolated* in A if, for some "root" vertex $a_0 \in VP$, we have

(8) $$E_0^P(a) = E_0^A(a) \quad \forall\, a \in VP - \{a_0\}.$$

It is then easily seen that P must be connected and, if $P \neq A$, then a_0 is unique. Moreover we have

(9) $$A = P \cup P^-, \quad P \cap P^- = \{a_0\},$$

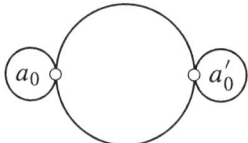

$Ends(A) = Ends(P) \amalg Ends(P^-)$, and P is determined, as a subgraph of A, by a_0 and $E_0^P(a_0)$ $(\subset E_0^A(a_0))$.

It follows that, since A is locally finite, the map $P \mapsto a_0$ (P an isolated subgraph $\neq A$) has finite fibers. Hence A has only countably many isolated subgraphs.

If P' is another isolated subgraph, with root $a_0' \neq a_0$,

then we have the following possible cases:

(10)
   (i) $a_0' \in P$:   Then either $P' \subset P$ or $P \cup P' = A$ (and $P^- \subset P'$).
               In the latter case, $Ends(A) = Ends(P) \cup Ends(P')$.
   (ii) $a_0' \in P^-$:   Then either $P \cap P' = \emptyset$ or $P \subset P'$.
           If $P \cap P' = \emptyset$ then $Ends(P) \cap Ends(P') = \emptyset$.

By a *parabolic tree* X we mean a tree with a unique end, $\varepsilon = end(X)$. These are studied in Chapter 9 below. One way to picture X is as a ray C to which one attaches rooted trees having no ends (hence finite trees if X is locally finite).

Another description (cf. (9.5)) is in terms of the sequence of "horospheres"

$$\cdots \to X_{n-1} \to X_n \to X_{n+1} \to \cdots \qquad (n \in \mathbf{Z}),$$

with

$$\varliminf X_n = \{\varepsilon\} \quad \text{and} \quad \varprojlim X_n = \emptyset.$$

The simplest parabolic tree is a ray, in which case we can take $X_n = \emptyset$ for $n < 0$, and $|X_n| = 1$ for $n \geq 0$. We call a parabolic tree *unbranched* if it is a ray.

By a *geometric parabolic end* of A we shall mean an end ε which is *the* end of an isolated parabolic subtree P of $A : \varepsilon = end(P)$. We claim:

(11)   Let $\varepsilon = end(P)$ and $\varepsilon' = end(P')$ be the ends of isolated parabolic subtrees P and P' of A, with distinct roots $a_0 \neq a_0'$. If $\varepsilon = \varepsilon'$, then one of P and P' contains the other, with finite complement. If $\varepsilon \neq \varepsilon'$, then $P \cap P'$ is finite. If $P \cap P' \neq \emptyset$, then $P \cup P' = A$, and $Ends(A) = \{\varepsilon, \varepsilon'\}$.

We use notation of (9) and (10). If say $P' \subset P$, then $\varepsilon = \varepsilon'$ and $(P - P') \cup \{a_0'\}$ is a (connected) tree without ends, hence finite. Similarly if $P \subset P'$. There remain, by (10), only the cases where $P \cap P' = \emptyset$, and $\varepsilon \neq \varepsilon'$, or $P \cup P' = A$ and $Ends(A) = \{\varepsilon, \varepsilon'\}$. If $\varepsilon \neq \varepsilon'$ then $P \cap P'$, having no ends, is finite. This concludes the proof of (11).

Put

$$GPE(A) = \{\text{geometric parabolic ends of } A\}.$$

It follows readily from (9), (10), and (11) that:

(12) $|GPE(A)| \leq \aleph_0$.
       $GPE(A)$ is a discrete subset of $Ends(A)$.
       If $GPE(A) = Ends(A)$, then $Ends(A)$ is finite.

In example (4.11) below we have $|GPE(A)| = 0 < 1 = |Ends(A)|$. On the other hand:

(13)   If $\pi_1(A)$ is finitely generated, e.g., if A is a tree, and if $Ends(A)$ is finite, then $GPE(A) = Ends(A)$.

If fact, using (7), we can reduce to the case when A is a tree. We can then choose $A' \subset A$ large enough (and finite) so that each component of $A - A'$ has a unique end. These components will then be isolated parabolic subtrees, representing each of the various ends of A.

We can say more:

(14)   If $\pi_1(A)$ is finitely generated, e.g., if A is a tree, and if $Ends(A)$ is infinite, then either
       (a) $|Ends(A)| = 2^{\aleph_0}$, or
       (b) $|Ends(A)| = \aleph_0$, and $GPE(A)$ is dense in $Ends(A)$.

Again, using (7), we reduce to the case when A is a tree. In view of (13), it suffices to show that:

$$[\exists \, \varepsilon \in Ends(A) - \overline{GPE(A)}] \quad \Rightarrow \quad |Ends(A)| = 2^{\aleph_0},$$

where the bar denotes closure. Choose $a_0 \epsilon VA$ and consider the ray $[a_0, \varepsilon)$, with vertex sequence $a_0, a_1, a_2, a_3, \ldots$. Then A consists of $[a_0, \varepsilon)$ with a rooted tree A_n attached to a_n for each $n \geq 0$.

(15)

Since $\varepsilon \notin GPE(A)$,

(16) A_n is infinite for infinitely many n.

Since $\varepsilon \notin \overline{GPE(A)}$,

(17) $GPE(A_n) = \emptyset$ for all $n \gg 0$.

After truncating an initial segment of (15) we see that it suffices to prove:

(18)   Assume that: $(*) A$ is infinite and $GPE(A) = \emptyset$. Then $|Ends(A)| = 2^{\aleph_0}$.

First note from (3) and (13) that

(19) $(*)$ $\Rightarrow$ $Ends(A)$ is infinite.

Choose $\varepsilon \epsilon Ends(A)$, and consider the representation (15) of A, now satisfying (16) and

(17') $GPE(A_n) = \emptyset$ $\forall n \geq 0$.

Thus infinitely many A_n satisfy $(*)$. By the same reasoning, each of these infinitely many A_n has a ray with infinitely many branches all satisfying $(*)$. This reasoning can be repeated indefinitely. It follows that we can build rays from a_0 in A all of which admit infinitely many multiple branching choices. There are thus $2^{\aleph_0}$ such rays, all representing distinct ends of A. We can summarize some of the conclusions above as follows.

(20)   Assume that $\pi_1(A)$ is finitely generated. Let

$$c = |GPE(A)| \quad \text{and} \quad c' = |Ends(A) - GPE(A)|.$$

(a) $0 \leq c \leq \aleph_0$, and either $0 \leq c' \leq \aleph_0$ or $c' = 2^{\aleph_0}$.
(b)   If $c < \aleph_0$ then $c' = 0$ or $c' = 2^{\aleph_0}$. If $c = \aleph_0$ then $c' > 0$.

We shall see in Theorem (4.13) that there are no further restrictions on (c, c'), even with $A = \Gamma \backslash X$, Γ a lattice on $X = X_d$, the d-regular tree, $d \geq 3$.

4.10 Γ-parabolic ends and Γ-cusps

Let Γ be a group acting on a tree X. As in (9.4)(5) below, we call the *action parabolic* if

(1)   (a) Γ fixes a unique end ε of X, and
        (b) Γ fixes no vertex of X.

Assume that Γ acts parabolically, as above. Since an inversion fixes no end, Γ acts without inversions. Let $x_0 \in VX$ and let the ray $[x_0, \epsilon)$ have vertex sequence $x_0, x_1, x_2, x_3, \ldots$. From (9.4)(1) and (2) we have:

(2)   (a) $\Gamma_{x_n} \leq \Gamma_{x_{n+1}}$ for all $n \geq 0$, with strict inclusion infinitely often. Put $\Gamma^0 = \bigcup_{n \geq 0} \Gamma_{x_n}$.

        (b) There is an exact sequence $1 \to \Gamma^0 \xrightarrow{incl} \Gamma \xrightarrow{\tau} \mathbf{Z}$ with $\ell(g) = |\tau(g)| \ \forall \ g \in G$ (notation of (5.1) below).

Let $x'_0 \in VX$, and say $[x'_0, \varepsilon)$ has vertex sequence $(x'_n)_{n \geq 0}$. Since $[x_0, \varepsilon)$ and $[x'_0, \varepsilon)$ are virtually equal, we have $x'_n = x_{n+n_0}$ for some integer n_0 and all $n \gg 0$. It follows that the ascending filtration $(\Gamma_{x_n})_{n \geq 0}$ of Γ^0 is, up to translation of indices and initial truncation, an invariant of the isomorphism class of the (parabolic) Γ-action. In particular, if $\Gamma \leq G = \mathrm{Aut}(X)$, then it is a G-conjugacy invariant. We shall say that the filtration $(\Gamma_{x_n})_{n \geq 0}$ is *virtually invariant*.

Assume further now that $\Gamma \leq G = \mathrm{Aut}(X)$. From (9.4)(6) below:

(3)   If Γ is discrete (X being locally finite), then $\Gamma = \Gamma^0$.

In fact, in view of (5.2) below, and (2) and (3) above,

(4)   A discrete group Γ acting (faithfully) without inversion acts parabolically iff Γ is infinite and locally finite.

Consider the quotient graph and graph of groups,

(5)
$$p : X \to A = \Gamma^0 \backslash X$$
$$\mathbf{A} = (A, \mathcal{A}) = \Gamma^0 \backslash\backslash X, \quad \text{and} \quad (A, i) = I(\mathbf{A}).$$

Then (cf. (9.4)(10)), A is a tree with an end ε_A such that each X-ray ending at ε projects to an A-ray ending at ε_A. Moreover, by (2)(a) above, the groups $\mathcal{A}_a$ increase as a approaches ε_A along a ray.

We call such an $(\mathbf{A}, \varepsilon_A)$ a *parabolic tree-of-groups*. This property is captured by (A, i, ε_A), as follows: For $e \in EA$ directed toward ε_A, $i(e) = 1$ and, along any ray toward ε_A, $i(\bar{e}) > 1$ for infinitely many e directed toward ε_A. Thus we can speak of (A, i, ε_A) being a *parabolic edge-indexed-tree*. These defining properties show that ε_A is uniquely determined by (A, i), and so we can even speak simply of (A, i) being a *parabolic edge-indexed-tree*. (Caution: After (4.9)(10) above, we defined a *parabolic tree* to be a tree with a unique end. The trees here on which parabolic trees-of-groups, and parabolic edge-indexed-trees are based need *not* be parabolic trees.)

Assume now that:

(6) X is a locally finite tree, $\Gamma \leq \text{Aut}(X)$ is discrete and acts without inversion.

An end ε of X is called a Γ-*parabolic end*, and Γ_ε is called a Γ-*parabolic subgroup*, if Γ_ε acts parabolically on X, i.e., Γ_ε fixes no end $\neq \varepsilon$, and Γ_ε fixes no vertex; equivalently Γ_ε is infinite and locally finite (cf. (4)). In fact, it follows from (4) and (5.2) below that Γ-parabolic subgroups are exactly the maximal infinite-locally finite subgroups of Γ. As in (2) above, they carry ascending filtrations that are virtually equivariant under G-conjugation.

We call ε a Γ-*cuspidal end*, and Γ_ε a Γ-*cuspidal subgroup*, if, further, it satisfies the condition:

(7)   If $x_0 \varepsilon VX$ and if $[x_0, \varepsilon)$ has vertex sequence $x_0, x_1, x_2, x_3, \ldots$, then $\Gamma_{x_n} \leq \Gamma_\varepsilon$ for all $n \gg 0$.

Note that ε and Γ_ε determine each other, and, if $g \epsilon \Gamma$, then $\varepsilon' = g\varepsilon$ is also a Γ-parabolic end, and $\Gamma_{\varepsilon'} = g\Gamma_\varepsilon g^{-1}$.

Let ε be a Γ-parabolic end, and C a ray toward ε. Then $|(\Gamma_\varepsilon)_x| \to \infty$ as $x \to \varepsilon$ along C, hence likewise for $|\Gamma_x|$. It follows that $p : C \to A$ is proper, so $p(C)$ has a unique end, which we denote $p(\varepsilon)$; it is clearly independent of C, representing ε.

Let ε' be another Γ-parabolic end, and C' is a ray toward ε'. If $g\epsilon\Gamma$ and $g\varepsilon = \varepsilon'$, the gC and C' are virtually equal, so $p(C) = p(gC)$ and $p(C')$ define the same end, $p(\varepsilon) = p(\varepsilon')$, of $A = \Gamma\backslash X$. We put

(8) $\Gamma - Cusps(X) = \{\Gamma\text{-cuspidal ends of } X\}/\Gamma$.

The class of ε is denoted $[\varepsilon]$.

We next show how to characterize this set in terms of (A, i). Call $P \subset A$ an *index-cuspidal subtree* of (A, i) if P is a subtree with an end ε_P such that, for any ray C in P toward ε_P, and for $e \epsilon EC$ directed toward ε_P, we have, $i(e) = 1$, and $i(\bar{e}) > 1$ for infinitely many such e. Clearly ε_P is uniquely determined by these conditions. Moreover any index-cuspidal ray in P ends at ε_P, and any two such rays share all but finitely many edges (they are *virtually equal*).

Let P' be another index cuspidal subtree of (A, i). We call P and P' (A, i)-*equivalent* if $P \cap P'$ contains an index-cuspidal ray. This implies that $\varepsilon_P = \varepsilon_{P'}$, and it defines an equivalence relation. We put

(9) $Cusps(A, i) \ (= Cusps(\mathbf{A}))$

       $= \{\text{index-cuspidal subtrees of } (A, i)\}/(A, i)\text{-equivalence}$.

The class of P is denoted $[P]$. We have a natural map

$$Cusps(A, i) \twoheadrightarrow Ends(A)$$

(10)

$$[P] \mapsto \varepsilon_P.$$

This can fail dramatically to be injective. In example (4.11) below, $|Cusps(A, i)| = 2^{\aleph_0}$ while $|Ends(A)| = 1$.

Now define

(11)
$$\Gamma - Cusps(X) \underset{\beta}{\overset{\alpha}{\rightleftarrows}} Cusps(A, i),$$

as follows. Let ε be a Γ-cuspidal end and C a ray in X toward ε. After removing an initial segment of C we can, by (7), assume that Γ_x increases (indefinitely) as $x \to \varepsilon$ along C. As observed above, $p(C)$ contains a ray P (toward $p(\varepsilon)$). Clearly then P is an index-cuspidal ray in (A, i). Moreover $[P]$ depends only on $[\varepsilon]$, clearly, so we can define $\alpha([\varepsilon]) = [P]$.

Conversely, let $P \subset A$ be an index-cuspidal subtree. Lift P to a tree $\tilde{P} \subset X$, and let ε be the end of $\tilde{P}$ over ε_P. It is clear then that ε is a Γ-cuspidal end of X. In fact if $C \subset P$ is an index-cuspidal ray, with lift $\tilde{C} \subset \tilde{P}$, then $\tilde{C}$ ends at ε and the stabilizers Γ_x increase indefinitely as $x \to \varepsilon$ along $\tilde{C}$. This shows further that $[\varepsilon]$ depends only on $[P]$, so we can put $\beta([P]) = [\varepsilon]$. Evidently,

(12) α and β are inverse bijections.

4.11   Unidirectional examples

Example 1. Let q be an integer ≥ 2 and put $d = 2(q + 1)$. Let

where, for $n \geq 1$,

$$i(e_0) = d \qquad \text{For } n \geq 1, \quad i(e_n) = 1 = i(f_n)$$
$$i(\bar{e}_0) = 2q \qquad\qquad\qquad\qquad i(\bar{e}_n) = q = i(\bar{f}_n),$$

(A, i) admits a faithful finite grouping

(2) $\mathbf{A} = (A, \mathcal{A})$

as follows. Let D, Q, T be groups of orders d, q, and 2, respectively. Then we put

$$\mathcal{A}_{a_0} = D, \quad \mathcal{A}_{e_0} = \{1\}, \quad \text{and}$$
$$\mathcal{A}_{a_n} = \mathcal{A}_{e_n} = \mathcal{A}_{f_n} = T \times Q^n \text{ for } n \geq 1,$$

with injections being the obvious inclusions. We have

(3)
$$\mathrm{Vol}(\mathbf{A}) = \frac{1}{d} + \frac{1}{2} \sum_{n \geq 1} \frac{1}{q^n} = \frac{1}{d} + \frac{1}{2(q-1)}$$
$$= \frac{1}{2} \left(\frac{1}{q+1} + \frac{1}{q-1} \right) = \frac{q}{q^2 - 1}.$$

Further

(4) $\pi_1(A)$ is not finitely generated, and

(5) A has a single end, ε, and ε is not a geometric parabolic end; moreover A has a single direction.

On the other hand, the maximal index-cuspidal subtrees of (A, i) are all rays of the form

$$C_{(u)} = (u_1, u_2, u_3, \ldots,) \ (u_n = e_n \text{ or } f_n \ \forall \ n \geq 1),$$

all ending at ε. Moreover $[C_{(u)}] = [C_{(u')}]$ iff $u_n = u'_n$ for all but finitely many n. It follows that:

(6) $|Cusps(A, i)| = 2^{\aleph_0}.$

Let

(7)
$$X = \widetilde{(A, i, a_0)} \text{ and}$$
$$\Gamma = \pi_1(\mathbf{A}, a_0) \leq \mathrm{Aut}(X).$$

Then

(8) $X = X_d$, the d-regular tree, and Γ is an X-lattice.

Moreover it follows from (6) and (4.10)(8) that

(9)   There are $2^{\aleph_0}$ Γ-conjugacy classes of Γ-cuspidal subgroups $\Gamma_\varepsilon \leq \Gamma$, ε a Γ-cuspidal end of X.

Example 2. For any $d \geq 3$, we can find a lattice Γ on $X = X_d$ such that $(A, i) = I(\Gamma \backslash\backslash X)$ has properties somewhat similar to those of Example 1; in particular, $\pi_1(A)$ is not finitely generated. More generally, we can make similar examples A_r so that $\pi_1(A_r)$ is a free group on r generators for any $r, 0 \leq r \leq \aleph_0$.

To this end, write $d = q + 1, q \geq 2$, and put

(1) $C = $

For $r = \aleph_0$, we take

(2) $(A_{\aleph_0}, i) = $

Clearly $\pi_1(A_{\aleph_0})$ is not finitely generated, and $(A_{\aleph_0}, i)$ has bounded denominators. Moreover,

(3)
$$\mathrm{Vol}_\bullet(A_{\aleph_0}, i) = 1 + d\left(\frac{1}{q} + \frac{2}{(q-1)q} + \frac{1}{q^2} + \frac{2}{(q-1)q^2} + \frac{1}{q^3} + \frac{2}{(q-1)q^3} + \cdots\right)$$
$$= 1 + d\left(\left(\sum_{n\geq 1}\frac{1}{q^n}\right) + \frac{2}{q-1}\left(\sum_{n\geq 1}\frac{1}{q^n}\right)\right)$$
$$= 1 + d\left(1 + \frac{2}{q-1}\right)\left(\frac{1}{q-1}\right) = 1 + \frac{d(q+1)}{(q-1)^2}$$

Let $\mathbf{A}$ be faithful finite grouping of $(A_{\aleph_0}, i)$ and $\Gamma = \pi_1(\mathbf{A}, \bullet)$. Then, just as in Example 1 (5),

(4) $A_{\aleph_0}$ has a single end ε_A, ε_A is not a geometric parabolic end, and $A_{\aleph_0}$ has a single direction. However, $A_{\aleph_0}$ has no index-cuspidal rays, so $\mathrm{Cusps}(A_{\aleph_0}, i) = \emptyset$.

Moreover, the grouping $\mathbf{A}$ can be chosen so that

(5)   There are $2^{\aleph_0}$ Γ-conjugacy classes of Γ-parabolic subgroups $\Gamma_\varepsilon \leq \Gamma$.

For example, choose groups D, Q', $Q_0 < Q_1 < Q_2 < \cdots$ with $|D| = d$, $|Q'| = q - 1$, and $|Q_n| = q^n$. From left to right on $A_{\aleph_0}$, put vertex groups D on the first vertex, and, on the nth copy of C, Q_n, $Q' \times Q_n$, $Q' \times Q_n$, Q_{n+1}. On the edges, put the obvious subgroups. Consider rays P in $A_{\aleph_0}$ from the initial vertex to ε_A. There are $2^{\aleph_0}$ of these. Fix $x_0 \in VX$ above the initial vertex, and lift P to a ray in X from x_0, with end ε ($= \varepsilon_P$). Then Γ_ε contains a copy of $Q = \cup_n Q_n$, so ε is parabolic. These ε fall into $2^{\aleph_0}$ Γ-orbits, whence (5).

For $r < \aleph_0$, we take

(6) $(A_r, i) = $

(r copies of) C

Clearly $\pi_1(A_r)$ is free on r generators. Moreover it is easily seen that (A_r, i) has bounded denominators, $\text{Vol}_\bullet(A_r, i) < \infty$, and (A_r, i) has a single end which is both an unbranched geometric parabolic end, and represents the unique Γ_r-cusp of X, where Γ_r denotes π_1 of a grouping $\mathbf{A}_r$ of (A_r, i).

If we replace the tail of (A_r, i) (following the copies of C) by the indexing used in the proof of (4.16) below, and also use the grouping constructed there, then we obtain a variant where A_r is unchanged, but the new l' has (as in (4.16)) no infinite, locally finite subgroups, hence there are no Γ-cusps.

4.12   A planar example

Let A be the Cayley graph of $\mathbf{Z}^2$ relative to $S = \{\pm x, \pm y\}$, where $x = (1, 0)$, and $y = (0, 1)$. Thus

(1)
$$VA = \mathbf{Z}^2 \text{ and}$$
$$EA = \{e = (z, z + s)|z = \partial_0 e \epsilon \mathbf{Z}^2, s \epsilon S, \partial_1 e = z + s\}.$$

We index A as follows. Fix integers $p, q \geq 2$. Consider edges $e = (z, z+s)$ "directed away from the origin," i.e., $z + s$ is farther from 0 (=(0,0)) than z. Then we put

(2)
$$i(e) = 1 \text{ and } i(\bar{e}) = \begin{cases} p & \text{if } s = \pm x \\ q & \text{if } s = \pm y. \end{cases}$$

(3) $(A_r, i) =$

Each "tile" in (A, i) has the form

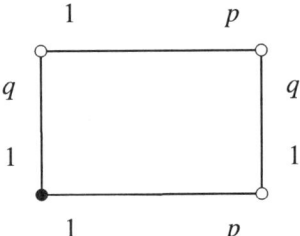

where the darkened vertex denotes the one nearest the origin 0. Hence (A, i) is unimodular. In fact (A, i) admits the faithful finite grouping

(4)
$$\mathbf{A} = (A, \mathcal{A}),$$
$$\mathcal{A}_{(m,n)} = P^{|m|} \times Q^{|n|},$$

where P and Q are groups of orders p and q, repectively. We then have

(5)
$$\mathrm{Vol}(\mathbf{A}) = \sum_{m,n \in \mathbf{Z}} \frac{1}{p^{|m|} q^{|n|}}$$
$$= V(p) \cdot V(q), \quad \text{where}$$

(6)
$$V(p) = \sum_{m \in \mathbf{Z}} \frac{1}{p^{|m|}}$$
$$-1 + 2 \sum_{m \geq 1} \frac{1}{p^m}$$
$$= 1 + \frac{2}{p-1} = \frac{p+1}{p-1}.$$

Put

(7)
$$X = \widetilde{(A, i, 0)} \quad \text{and}$$
$$\Gamma = \pi_1(\mathbf{A}, 0) \leq \mathrm{Aut}(X), \quad \text{a non-uniform } X\text{-lattice.}$$

The tree X is not regular, but has bounded degree. In fact:
For $v \in VX$ and $p : X \to A$,

(8)
$$\deg(v) = \begin{cases} 4 & \text{if } p(v) = 0 \\ p+3 & \text{if } p(v) = (m, 0) \neq 0 \\ q+3 & \text{if } p(v) = (0, n) \neq 0 \\ p+q+2 & \text{else.} \end{cases}$$

(9)   Put $d = p + q + 2$. For $z = (m, n) \in \mathbf{Z}^2$, put $r(z) = d - \deg_{(A,i)}(z)$.

Then, as in (4.1)(4), we can graft a copy of $B(r(z), d)$, to (A, i) at z, where

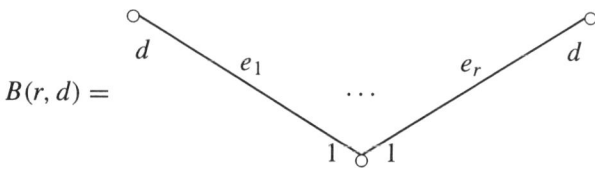

$$B(r, d) =$$

This enlarges (A, i) to (A^+, i), and we extend $\mathbf{A}$ to a grouping $\mathbf{A}^+$ of (A^+, i) by putting the group $\mathcal{A}_z \times D$ (D a fixed group of order d) at the new vertices adjacent to z. Then we have:

(10) $X^+ = \widehat{(A^+, i, 0)} = X_d$, the d-regular tree, $\Gamma^+ = \pi_1(\mathbf{A}^+, 0)$ is an X^+-lattice with $\Gamma^+ \backslash\backslash X^+ = \mathbf{A}^+$, as above, and

$$\text{Vol}(\Gamma^+ \backslash\backslash X^+) = \frac{(p+1)(q+1)}{(p-1)(q-1)} + \frac{1}{d}\left(d - 4 + 2\left(\frac{p-1}{q-1} + \frac{q-1}{p-1}\right)\right).$$

The volume formula is a straightforward calculation. As for the geometry of A^+:

(11)
$$\pi_1(A^+) \text{ is not finitely generated,}$$
$$A^+ \text{ has 1 end,}$$
$$A^+ \text{ has } 2^{\aleph_0} \text{ directions, and}$$
$$A^+ \text{ has no geometric parabolic ends.}$$

On the other hand,

(12)   All rays that move steadily away from the origin in A are index-cuspidal rays of (A, i) and of (A^+, i). Two such define the same cusp iff they are "virtually equal." Thus $|Cusps(A^+, i)| \ (= |\Gamma^+\text{-}Cusps(X^+)|) = 2^{\aleph_0}$.

4.13   Theorem. *Let $X = X_d$, the d regular tree, $d \geq 3$. Let c, c' be cardinals such that:*

(1) $0 \leq c \leq \aleph_0$ *and either* $0 \leq c' \leq \aleph_0$ *or* $c' = 2^{\aleph_0}$.

(2)
If $c < \aleph_0$ then either $c' = 0$ or $c' = 2^{\aleph_0}$.
If $c = \aleph_0$ then $c' > 0$.

Then there is an X-lattice $\Gamma = \Gamma_{c,c'}$, such that $A = A_{c,c'} := \Gamma\backslash X$ is a tree satisfying:
(a) *$|GPE(A)| = c$, and every geometric parabolic end of A is unbranched.*
(b) *$|Ends(A) - GPE(A)| = c'$.*

(c) *The map Cusps(A,i) → Ends(A), [P] ↦ ε_P* ((4.10)(10)) *is bijective.*

4.14 **Notation and remark.** For any connected locally finite graph A let us write

(1)
$$c(A) = |GPE(A)|, \quad \text{and}$$
$$c'(A) = |Ends(A) - GPE(A)|.$$

We have seen in (4.9)(20) that, if $\pi_1(A)$ is finitely generated, then $c = c(A)$ and $c' = c'(A)$ satisfy conditions (1) and (2) of (4.13). Theorem (4.13) shows therefore that there are no further constraints on (c, c').

4.15 **Proof of (4.13).** We shall construct an edge indexed tree

$$(A, i) \ (= (A_{c,c'}, i))$$

such that

(1) (A, i) satisfies the conclusions of (4.13);

(2) (A, i) is d-regular, i.e., $\sum_{e \in E_0(a)} i(e) = d \ \forall \ a \in VA$;

(3) (A, i) has "bounded denominators," in the sense of (2.6)(18); and

(4) $Vol(A, i) < \infty$ in the sense of (2.6)(13).

Given this, choose $a_0 \in VA$. It follows then from (2) that

$$X = \widehat{(A, i, a_0)} = X_d$$

From (3) and (2.6)(19), we can find a faithful finite grouping $\mathbf{A} \ (= \mathbf{A}_{c,c'})$ of (A, i), and hence $\Gamma \ (= \Gamma_{c,c'}) = \pi_1(\mathbf{A}, a_0) \le Aut(X)$ with $\Gamma \backslash\backslash X = \mathbf{A}$. Finally, (4) guarantees that Γ is an X-lattice. Thus it suffices to construct (A, i) satisfying (1)–(4). In doing this it will be convenient to write

(5)
$$d = q + 1, \quad q \ge 2.$$

Our constructions will have the following forms, with notations C', B_c, and $D_{c'}$ to be explained below.

(6)   For $c = 0$,

$$(A_{0,0}, i) = \overset{d \qquad d}{\underset{\circ\rule{3em}{0.4pt}\circ}{}}, \quad \text{and}$$

$$(A_{0,2^{\aleph_0}}, i) = \overset{C' \qquad\qquad q \qquad d}{\underset{\rule{4em}{0.4pt}\circ\rule{3em}{0.4pt}\circ}{}}.$$

(7)   For $0 < c < \aleph_0$,

$$(A_{c,0}, i) = \underset{\circ}{\overset{d}{\rule{2cm}{0.4pt}}}\overset{q}{\rule{2cm}{0.4pt}}\underset{\circ}{}\overset{B_c}{\rule{2cm}{0.4pt}},$$

and, for $0 < c \le \aleph_0$,

$$(A_{c,2\aleph_0}, i) = \overset{C'}{\rule{2cm}{0.4pt}}\underset{\circ}{}\overset{q}{\rule{2cm}{0.4pt}}\overset{q}{\rule{2cm}{0.4pt}}\underset{\circ}{}\overset{B_c}{\rule{2cm}{0.4pt}}$$

(8)   For $c = \aleph_0$ and $0 < c' < \aleph_0$,

$$(A_{\aleph_0,c'}, i) = \underset{\circ}{\overset{d}{\rule{2cm}{0.4pt}}}\overset{q}{\rule{2cm}{0.4pt}}\underset{\circ}{}\overset{D_{c'}}{\rule{2cm}{0.4pt}}.$$

Notation:

(9)   For $E = C'$, B_c $(0 < c \le \aleph_0)$, or $D_{c'}$ $(0 \le c' \le \aleph_0)$, to be constructed below:
   (a) E is an edge-indexed tree, attached in (6)–(8) above at a root vertex $\bullet$.
   (b) $\deg_E(\bullet) = 1$, and, for $a \in VE - \{\bullet\}$, $\deg_E(a) = d$, and a is not a geometric terminal vertex of E.
   (c) $\mathrm{Vol}_\bullet(E) < \infty$.
   (d) All rays from $\bullet$ in E are index-cuspidal rays of E, in the sense of (4.10).

Further we have:

(10)
$$\begin{aligned}
c(B_c) &= c, & c'(B_c) &= \begin{cases} 0, & c < \aleph_0, \\ 1, & c = \aleph_0, \end{cases} \\
c(D_{c'}) &= \aleph_0, & c'(D_{c'}) &= c', \\
c(C') &= 0, & c'(C') &= 2^{\aleph_0}.
\end{aligned}$$

Now requirement (2) follows from (9)(b), (3) follows from (9)(a) and (d), and (4) follows from (9)(c). As for the assertions of Theorem (4.13), (4.13)(a) follows from (6)–(8) and (10), and (9)(b) (for the "unbranched" assertion), (4.13)(b) follows from (6)–(8) and (10), and (4.13)(c) follows from (9)(a) and (d).

It remains now to construct the C', B_c, $D_{c'}$ satisfying (9) and (10). We start with:

(11) $C = \overset{\bullet}{}\overset{1}{\rule{1.5cm}{0.4pt}}\underset{\circ}{}\overset{q}{\rule{1.5cm}{0.4pt}}\overset{1}{\rule{1cm}{0.4pt}}\underset{\circ}{}\overset{q}{\rule{1.5cm}{0.4pt}}\overset{1}{\rule{1cm}{0.4pt}}\underset{\circ}{}\overset{q}{\rule{1.5cm}{0.4pt}}\overset{1}{\rule{1cm}{0.4pt}}\ \cdots$

$$c(C) = 1, \quad c'(C) = 0, \quad \text{and} \quad \mathrm{Vol}_\bullet(C) = \sum_{n \ge 0} \frac{1}{q^n} = \frac{q}{q-1}.$$

(12)   Put $B_1 = C$, and, for $1 < c < \aleph_0$,

$$B_c = \underset{1}{\overset{\bullet}{\rule{0pt}{0pt}}}\ \overset{C}{\underset{q-1}{\rule{2cm}{0.4pt}}}\underset{\circ}{}\ \overset{C}{\underset{1\quad q-1}{\rule{2cm}{0.4pt}}}\underset{\circ}{}\ \overset{C}{\underset{1\quad q-1}{\rule{1cm}{0.4pt}}}\cdots\overset{C}{\underset{1\quad q-1}{\rule{2cm}{0.4pt}}}\underset{\circ}{}\ \overset{C}{\underset{1\quad q-1}{\rule{2cm}{0.4pt}}}\underset{\circ}{}\ \overset{C}{\underset{1\quad q-1}{\rule{2cm}{0.4pt}}}\overset{C}{}$$

(c copies of C).

Then $c(B_c) = c$, $c'(B_c) = 0$, and

$$\text{Vol}_\bullet(B_c) = 1 + \frac{q}{(q-1)^2} + \frac{q}{(q-1)^3} + \cdots + \frac{q}{(q-1)^{c-1}} + \frac{2q}{(q-1)^c}.$$

(13)   Define $\beta = B_{\aleph_0} :=$

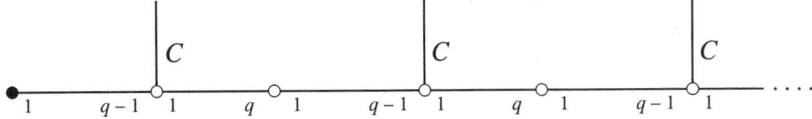

We have $c(\beta) = \aleph_0$, $c'(\beta) = 1$ (the horizontal ray), and, putting

$$v = \text{Vol}_\bullet(C) = \frac{q}{q-1} \quad \text{(cf. (11))},$$

we have

$$\text{Vol}_\bullet(\beta) = 1 + \frac{v}{q-1} + \frac{1}{(q-1)q} + \frac{v}{(q-1)^2 q} + \frac{1}{(q-1)^2 q^2} + \cdots$$

$$= \sum_{n \geq 0} \left(\frac{1}{(q-1)q}\right)^n + \frac{v}{q-1}\left(\sum_{n \geq 0}\left(\frac{1}{(q-1)q}\right)^n\right)$$

$$= \left(1 + \frac{v}{q-1}\right)\left(\frac{1}{1 - \frac{1}{(q-1)q}}\right).$$

Next we introduce $D_{c'}$ $(0 < c' \leq \aleph_0)$:

(14)   For $0 < c' < \aleph_0$,

$$D_{c'} = \qquad \qquad \qquad \qquad \qquad \qquad \qquad$$

(c' copies of β)

Then $c(D_{c'}) = \aleph_0$ and $c'(D_{c'}) = c'$.

$$\text{Vol}_\bullet(D_{c'}) = 1 + \text{Vol}(\beta)\left(\frac{1}{q-1} + \frac{1}{(q-1)^2} + \cdots + \frac{2}{(q-1)^{c'-1}}\right).$$

(15) $D_{\aleph_0} =$

Then $c(D_{\aleph_0}) = \aleph_0$ and $c'(D_{\aleph_0}) = \aleph_0$.

$$\mathrm{Vol}(D_{\aleph_0}) = \left(\sum_{n \geq 0} \frac{1}{((q-1)q)^n} \right) \left(1 + \frac{\mathrm{Vol}(\beta)}{q-1} \right) < \infty.$$

Finally:

(16)   Fix an integer $r \geq 3$, put

$$R = \underset{1}{\bullet} \overset{q}{\underset{1}{\circ}} \overset{q}{\underset{1}{\circ}} \cdots \overset{q}{\underset{1}{\circ}} \overset{q}{\underset{1}{\circ}} \overset{q-1}{\circ} \qquad (r \text{ edges}),$$

and

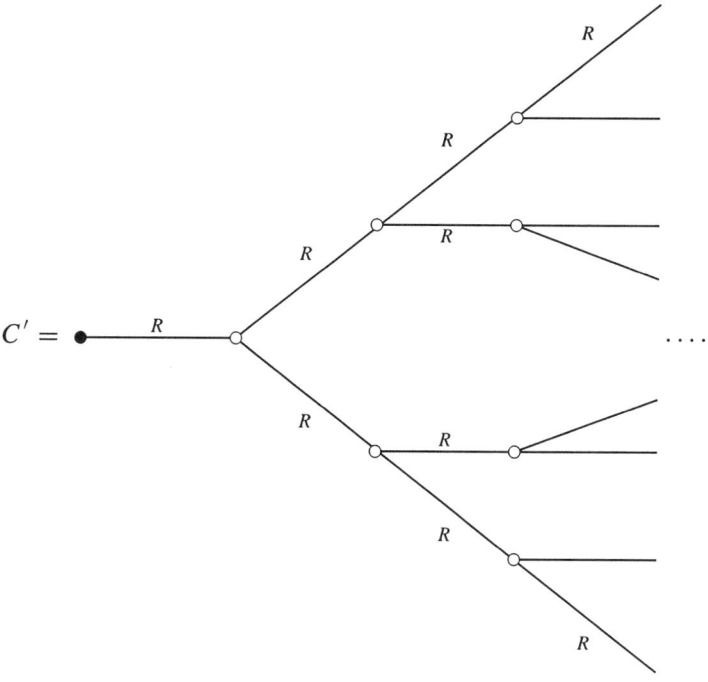

Clearly, then $c(C') = 0$, and $c'(C') = 2^{\aleph_0}$. We have $\mathrm{Vol}_\bullet(R) = w + \frac{1}{q^{r-1}(q-1)} = \frac{q}{q-1}$, where

$$w = 1 + \frac{1}{q} + \cdots + \frac{1}{q^{r-1}} = \frac{q^r - 1}{q^{r-1}(q-1)}$$

and

$$\mathrm{Vol}_\bullet(C') = w + \frac{2}{q^{r-1}(q-1)} \left(w + \frac{2}{q^{r-1}(q-1)} \left(w + \frac{2}{q^{r-1}(q-1)} \left(\cdots \right. \right. \right.$$

$$= w \left(1 + \sum_{n \geq 1} \left(\frac{2}{q^{r-1}(q-1)} \right)^n \right) < \infty$$

because $q \geq 2$ and $r \geq 3$.

The above constructions complete the proof of (4.13). With these ingredients the required $(A_{c,c'}, i)$ are given by (6)–(8).

All of the non-uniform lattices Γ exhibited so far have had Γ-parabolic subgroups ((4.10)(6)). We now show that this is not inevitable.

4.16 **Theorem.** *Let $X = X_d$, the d-regular tree, $d \geq 3$. Then there is an X-lattice Γ such that*

$$\Gamma \backslash X = \circ\!\!-\!\!-\!\!-\!\!-\circ\!\!-\!\!-\!\!-\circ\!\!-\!\!-\!\!-\circ\!\!-\!\!-\!\!-\circ\!\!-\!\!-\!\!-\cdots,$$

(hence $c(\Gamma \backslash X) = 1$ and $c'(\Gamma \backslash X) = 0$), and Γ has no infinite, locally finite subgroups. Hence there are no Γ-parabolic ends, in particular no Γ-cusps.

Proof. We start by building $(A, i) = I(\Gamma \backslash\!\backslash X)$. Write $d = q + 1$, $q \geq 2$, and choose an integer $t \geq 1$. Put

(1) $\quad U = \bullet\!\!\underset{}{\overset{1}{-\!\!-\!\!-}}\!\!\underset{}{\overset{q}{\circ}}\!\!\overset{1}{-\!\!-}\!\!\underset{}{\overset{q}{\circ}}\!\!\overset{1}{-\!\!-}\cdots\overset{q}{-\!\!-}\!\!\underset{}{\overset{1}{\circ}}\!\!\overset{q}{-\!\!-}\circ \qquad$ (t edges)

and

(2) $\qquad\qquad D = \bullet\!\!\underset{}{\overset{1}{-\!\!-\!\!-}}\!\!\underset{}{\overset{1}{\circ}}\!\!\overset{q}{-\!\!-}\!\!\underset{}{\overset{q}{\circ}}.$

Then we take

(3) $\quad (A, i) =$

$$\bullet\!\!\overset{d}{\underset{\longrightarrow}{-\!\!-\!\!-}}\!\!\overset{q}{\circ}\!\!\overset{U}{-\!\!-}\!\!\circ\!\!\overset{D}{-\!\!-}\!\!\circ\!\!\overset{U}{-\!\!-}\!\!\circ\!\!\overset{D}{\cdots-\!\!-}\!\!\overset{U}{\circ}\!\!\overset{D}{-\!\!-}\!\!\circ\!\!\overset{U}{-\!\!-}\cdots\cdot$$

We have

(4)
$$\mathrm{Vol}_\bullet(U) = u + \frac{1}{q^t}, \quad \text{where}$$
$$u = 1 + \frac{1}{q} + \cdots + \frac{1}{q^{t-1}} = (q^t - 1)/q^{t-1}(q - 1), \quad \text{and}$$
$$\mathrm{Vol}_\bullet(D) = 3.$$

Thus

(5)
$$\mathrm{Vol}_\bullet(A, i) = 1 + \frac{1}{q}\left(u + \frac{1}{q^t}\left(2 + u + \frac{1}{q^t}\left(2 + u + \frac{1}{q^t}(2 + u + \cdots)\right)\right)\right)\cdots$$
$$= 1 + \frac{1}{q}\left(u + (2 + u)\left(\sum_{n \geq 1}\frac{1}{q^{tn}}\right)\right) < \infty.$$

To construct a faithful finite grouping $\mathbf{A} = (A, \mathcal{A})$ of (A, i) with the desired properties we introduce the following notations. Fix a group Q of order $|Q| = q$. In the infinite product $P = Q^{\mathbf{N}}$ we view the subproducts $Q^V (V \subset \mathbf{N})$ as subgroups of P. If $V \subset V'$ then $Q^V \leq Q^{V'}$.

(6)   For $V \subset \mathbf{N}$ we have $V + 1 \subset \mathbf{N}$, and put $V^+ = V \cup (V + 1)$ and $V^- = V \cap (V + 1)$.

For an interval $[a, b]$, $a \leq b$, in $\mathbf{N}$,

(7) $$[a, b]^+ = [a, b + 1], \quad [a, b]^- = [a + 1, b].$$

Now list the vertices and edges of A as follows.

We start with:

(8) $$\mathcal{A}_0 = \text{any group of order } d.$$
$$\mathcal{A}_{e_1} = \{1\}.$$

For $n \geq 1$ we define

(9) $$A_n = Q^{V(n)} \quad \text{and} \quad A_{e_{n+1}} = Q^{E(n+1)}$$

inductively, as follows, starting with

$$E(1) = \emptyset \text{ and } V(1) = \{1\} = E(2) \quad (\text{so } \mathcal{A}_1 \cong Q \cong \mathcal{A}_{e_2}).$$

Suppose that $n > 1$ and that we have defined $V(r)$ and $E(r)$ for $r < n$.

There are three possible cases.

(10) $$(i(e_n), i(\bar{e}_n)) = \begin{cases} \text{(i)} & (1, q) \\ \text{(ii)} & (1, 1) \\ \text{(iii)} & (q, q) \end{cases}$$

We now define $V(n)$ and $E(n)$ in each case.

(11) $$\begin{cases} \text{case (i):} & E(n) = V(n - 1), \text{ and } V(n) = E(n)^+ \quad (\text{notation of (6)}) \\ \text{case (ii):} & E(n) = V(n - 1) = V(n) \\ \text{case (iii):} & E(n) = V(n - 1)^-, \text{ and } V(n) = E(n)^+ \quad (\text{notation of (6)}) \end{cases}$$

It follows inductively that,

(12) $A_n = Q^{V(n)}$ where $V(n)$ is an interval $[a(n), b(n)]$ in $\mathbf{N}$,

$$|A_n| = q^{(b(n)-a(n)+1)} \to \infty,$$

and $a(n) \to \infty$ as $n \to \infty$.

In fact, $a(1) = 1 = b(1)$, and from (7) and (11) we have

case (i) : $a(n) = a(n-1),$ $b(n) = b(n-1) + 1$
case (ii) : $a(n) = a(n-1),$ $b(n) = b(n-1),$
case (iii) : $a(n) = a(n-1) + 1,$ $b(n) = b(n-1) + 1,$

and each case occurs infinitely often. It follows from (12) that

(13)   For $n \geq 1$, there is an $r(n) \geq 1$ such that $V(n) \cap V(n+r) = \emptyset$ for $r \geq r(n)$.

Now let

(14)
$$X = \widehat{(A, i, \bullet)} \cong X_d, \quad \text{and}$$
$$\Gamma = \pi_1(\mathbf{A}, \bullet) \leq \text{Aut}(X).$$

Then, in view of (5), Γ is an X-lattice with $\Gamma \backslash\backslash X = \mathbf{A}$.

The group Γ is the free product of the groups A_n with the edge groups amalgamated. For $n \geq 1$, all vertex and edge groups are of the form $Q^V \leq Q^{\mathbf{N}} = P$, and the homomorphisms from edge groups to end point vertex groups are inclusions inside P. It follows that there is a homomorphism

(15)
$$\varphi : \Gamma \to P = Q^{\mathbf{N}}$$
$$\varphi(A_0) = \{1\}, \quad \text{and, for } n \geq 1,$$
$$\varphi|A_n = Q^{V(n)} \text{ is the inclusion.}$$

In particular, in view of (13),

(16)   For $n \geq 1$, $\varphi|A_n$ is injective, $\varphi(A_n) \triangleleft P$, and there is an $r(n) \geq 1$ such that $\varphi(A_n) \cap \varphi(A_{n+r}) = \{1\}$ for $r \geq r(n)$.

We now conclude the proof of Theorem (4.16) by showing that Γ cannot contain an infinite, locally finite group. In fact let L be such a group, and let

$$p : X \to A = \Gamma \backslash X$$

be the projection. According to (5.2) below, L fixes a (unique) end ε of X. Let $x_0, x_1, x_2, x_3, \ldots$ be the vertex sequence of a ray in X ending at ε. Then (loc. cit.) L is the ascending union of the (finite) stabilizers L_{x_k}. Putting $n(k) = p(x_k)$, L_{x_k} is conjugate to a subgroup of $A_{n(k)}$. Since $|L_{x_k}| \to \infty$ as $k \to \infty$, we must have $|A_{n(k)}| \to \infty$, hence also $n(k) \to \infty$, as $k \to \infty$.

Choose k so that $L_{x_k} \neq \{1\}$ and $n(k) \geq 1$. Choose $\ell > k$ so that $n(\ell) \geq n(k) + r(n(k))$, with the notation of (16). Then we have

(17) $\{1\} \neq L_{x_k} \leq L_{x_\ell}$.

L_{x_k} is conjugate to a subgroup of $\mathcal{A}_{n(k)}$.

L_{x_ℓ} is conjugate to a subgroup of $\mathcal{A}_{n(\ell)}$; $n(\ell) \geq n(k) + r(n(k))$.

It follows from (17) that there is a $g \in \Gamma$ such that

(18) $\mathcal{A}_{n(k)} \cap g \mathcal{A}_{n(\ell)} g^{-1} \neq \{1\}$.

Applying φ (in (15)) to (18), and using (16), we have

(19) $Q^{V(n(k))} \cap Q^{V(n(\ell))} \neq \{1\}$ in P.

Since $n(\ell) \geq n(k) + r(n(k))$ we have $V(n(k)) \cap V(n(\ell)) = \emptyset$, by (13), so (19) is impossible. This completes the proof of Theorem (4.16).

Finally, we show that, if one is not concerned about regularity, or bounded degree, of X, then every possible graph is a lattice quotient.

4.17   Theorem (All Quotients Theorem). *Let A be any connected locally finite graph. Then there is a locally finite tree X and an X-lattice Γ such that $\Gamma \backslash X \cong A$.*

Proof. In view of (2.6)(19) and (15), it suffices to produce an edge-indexing (A, i) such that (A, i) is unimodular with bounded denominators, and $\mathrm{Vol}(A, i) < \infty$.

Choose a maximal tree $T \subset A$ and a base point $a_0 \in VA = VT$. In T let

(1) $T_n = \{a \in VT \mid dist_T(a, a_0) = n\}$,

the (finite) n-sphere in T centered at a_0. For $a \in T_n$ there is a unique $e \in E_0(a)$ such that $t(a) := \partial_1(e) \in T_{n-1}$. The rooted tree (T, a_0) is in fact completely determined by the sphere sequence.

(2) $\{a_0\} = T_0 \overset{t}{\longleftarrow} T_1 \overset{t}{\longleftarrow} T_2 \leftarrow \cdots \leftarrow T_{n-1} \overset{t}{\longleftarrow} T_n \leftarrow \cdots$.

We have

$$ET = E^+T \sqcup \overline{E^+T}, \quad \text{where}$$
$$E^+T = \{e \in ET \mid dist_T(\partial_0 e, a_0) < dist_T(\partial_1 e, a_0)\}$$
$$= \sqcup_{n \geq 1} E_n^+ T, \quad \text{where} \quad E_n^+ T = \{e \in ET \mid \partial_0 e \in T_{n-1}, \partial_1 e \in T_n\}.$$

For each $n \geq 1$ choose an integer

$$q_n > |T_n|.$$

For $e \in E_n^+ T$, put $i(e) = 1$ and $i(\bar{e}) = q_n$.

Then, relative to (T, i), and the base point a_0, we have, for $a \in T_n$,

$$N_{a_0}(a) = \frac{\Delta a}{\Delta a_0} = m_n := q_1 q_2 \ldots q_n \in \mathbf{N}.$$

It follows that

(3) (T, i) has bounded denominators, and

(4)
$$\mathrm{Vol}_{a_0}(T, i) = \sum_{n \geq 0} \frac{|T_n|}{m_n} = 1 + \sum_{n \geq 1} \frac{1}{m_{n-1}} \left(\frac{|T_n|}{q_n} \right)$$

$$< 1 + \sum_{n \geq 1} \frac{1}{m_{n-1}} \leq 1 + \sum_{n \geq 0} \frac{1}{2^n} < \infty.$$

It suffices therefore to show that we can complete the indexing i to all of EA so that (A, i) is unimodular. To do this, consider any $f \in EA$. At the cost of replacing f by $\bar{f}$, if necessary, we can assume that

(5) $\partial_0 f \in T_r, \quad \partial_1 f \in T_s, \quad$ and $r \leq s$.

Then we put

(6) $i(f) = 1$ and $i(\bar{f}) = \dfrac{m_s}{m_r} = q_{r+1} \cdot q_{r+2} \ldots q_s.$

It is then easy to see that the resulting (A, i) is unimodular, as required.

Chapter 5

Length Functions, Minimality

We review here for reference the properties of hyperbolic length (of tree automorphisms) and the information it carries about tree actions. We fix a tree X, $G = \text{Aut}(X)$, and let d ($= d_X$) denote the edge-path distance on VX.

5.1 Hyperbolic length (cf. [B3], II, §6)

For $s \in G$, we define $\ell(s) \in \mathbf{Z}$ and a subtree $X_s \subset X$ as follows.

(1)   If s inverts an edge e, $se = \bar{e}$, then we put $\ell(s) = 0$, $X_s = \circ\!\!\overset{e}{\rule{1cm}{0.4pt}}\!\!\circ$, and call s an *inversion*. Every $\langle s \rangle$-invariant subtree contains X_s. (It is more natural here to take X_s to be the "midpoint" of e, a vertex not of X but of its barycentric subdivision, where it is the unique fixed point of s.)

(2)   If s is not an inversion, then

$$\ell(s) = \text{Min}_{x \in VX}\, d(sx, x), \quad \text{and} \quad VX_s = \{x \in VX \mid d(sx, x) = \ell(s)\}.$$

(3)   If $\ell(s) = 0$ and s is not an inversion, then X_s is the tree of fixed points of s; we call s *elliptic* in this case. Every $\langle s \rangle$-invariant subtree meets X_s.

(4)   If $\ell(s) > 0$, then X_s is a linear tree on which s induces a translation of amplitude $\ell(s)$; we then call s *hyperbolic*, and call X_s the "s-axis." Any $\langle s \rangle$-invariant subtree contains X_s.

(5)   (a) If $e \in EX$, $se \neq e$, and both e and se belong to a reduced path in X, then s is hyperbolic, $e \in EX_s$, and $\ell(s) = d(\partial_0 e, \partial_0 se)$. Consequently:

   (b) If $s, t \in G$, $\ell(s) > 0$ and $te = se$ for some $e \in EX_s$, then $\ell(t) = \ell(s) > 0$, and $e \in EX_t$.

(6)   For $s \in G$ and $x \in VX$, $d(sx, x) (\text{mod}\, 2) \in \mathbf{Z}/2\mathbf{Z}$ is independent of x; denote this $\bar{\ell}(s)$. Then $\bar{\ell} : G \to \mathbf{Z}/2\mathbf{Z}$ is a homomorphism. If s is not an inversion then

$\bar{\ell}(s) = \ell(s)(\text{mod}2)$, while $\bar{\ell}(s) = 1$ if s is an inversion. Thus $G^0 := Ker(\bar{\ell})$ is a subgroup of G without inversions, and of index ≤ 2.

(7)   For $s, t \epsilon G$, $\ell(tst^{-1}) = \ell(s)$ and $X_{tst^{-1}} = tX_s$. For $n \epsilon \mathbf{Z}$, $\ell(s^n) = |n|\ell(s)$, and $X_s \subset X_{s^n}$, with equality if $n \cdot \ell(s) \neq 0$. Hence $\ell(s) = 0$ if s has finite order. If $Y \subset X$ is an $\langle s \rangle$-invariant subtree, then $\ell(s|Y) = \ell(s)$.

(8)   An action of a group Γ on X is given by a homomorphism $\rho : \Gamma \to G$. We call X, with this action, a Γ-*tree*. And we write $\text{Aut}_\Gamma(X) = Z_G(\rho\Gamma)$, the group of Γ-equivariant automorphisms of X. We denote $\ell \circ \rho$ simply by ℓ, when ρ is understood, and call

$$\ell : \Gamma \to \mathbf{Z}$$

the "*hyperbolic length function*" of the Γ-tree X.

By (7),

(9) $\ell(\Gamma) = 0$ if Γ is torsion.

5.2   Proposition (cf. [AB], (3.4) or [T], (7.5).) *Suppose that*

$$\ell(\Gamma) = 0.$$

Then exactly one of the following occurs:
  (a) Γ *fixes some* $x \epsilon VX$.
  (b) Γ *inverts some* $e \epsilon EX : \Gamma \cdot e = \{e, \bar{e}\}$
  (c) Γ *fixes a (unique) end,* ε, *but no vertex of* X.
  In case (c), *if* $x_0 \epsilon VX$ *and if* $x_0, x_1, x_2, \ldots$ *is the vertex sequence of the ray* $[x_0, \varepsilon)$ *from* x_0 *to* ε, *then* Γ *is the infinitely ascending union of the vertex stabilizers* Γ_{x_n} ($n = 0, 1, 2, 3, \ldots$). *In particular,* Γ *is not finitely generated.*

5.3   Corollary. *Suppose that* Γ *is finitely generated and* $\ell(\Gamma) = 0$. *Then* Γ *either fixes a vertex or inverts an edge. If* Γ *is discrete then* Γ *is finite.*

5.4   Minimality

We call the action of Γ on X *minimal,* and call X a *minimal* Γ-*tree,* if there are no proper Γ-invariant subtrees. In the setting of (5.2)(c), the "horoballs" centered at ε (cf. (9.2) below) will be Γ-invariant subtrees.

In case X has terminal vertices (of degree 1), then by removing these and the incident edges, one is left with a proper subgraph X' of X which is invariant under $G = \text{Aut}(X)$. Moreover, $X' \neq \emptyset$, except in the case $X = \circ\!\!-\!\!-\!\!\circ$.

5.5   Corollary. *If* Γ *acts minimally and* $\ell(\Gamma) = 0$, *then* $X = \circ$ *(a vertex) or* $X = \circ\!\!-\!\!\!-\!\!\circ$ *(an edge).*

5.6   Proposition (cf. [B3], (7.9)).
(a) *If* $\Gamma\backslash X$ *is finite and each* Γ_x *is finitely generated, then* Γ *is finitely generated.*
(b) *If* Γ *is finitely generated and acts minimally on* X, *then* $\Gamma\backslash X$ *is finite.*

5.7   Proposition ([CM], (3.1) and [B3]). *If* $\ell(\Gamma) \neq 0$, *then there is a unique minimal* Γ*-invariant subtree* X_Γ, *and*

$$X_\Gamma = \bigcup_{s\in\Gamma, \ell(s)>0} X_s.$$

Moreover, X_Γ *is invariant under* $N_G(\Gamma)$, *and even* $C_G(\Gamma)$ (*cf.* (6.1)(8)).

5.8   Proposition. *Let* $\Gamma_0, \Gamma_1 \leq \Gamma$ *and* $t\in\Gamma$, *and assume that* $\ell(\Gamma_0) \neq 0$.
(a) $tX_{\Gamma_0} = X_{t\Gamma_0 t^{-1}}$.
(b) *If* $\Gamma_0 \triangleleft \Gamma$ *then* $X_{\Gamma_0} = X_\Gamma$.
(c) *If* Γ_0 *and* Γ_1 *are commensurable, or if* $\Gamma_0\backslash X = \Gamma_1\backslash X$, *then* $\ell(\Gamma_1) \neq 0$ *and* $X_{\Gamma_1} = X_{\Gamma_0}$.

5.9   Proposition (cf. [B3], (8.6)(b)). *Suppose that* Γ *acts minimally without inversion on* X *and* $\{1\} \neq N \triangleleft \Gamma$ *is a finitely generated normal subgroup. Then* N *acts minimally also, and* $N\backslash X$ *is finite. If, further,* Γ *is discrete, then* N *and* Γ *are uniform* X*-lattices, and* Γ/N *is finite.*

Proof. If $\ell(N) \neq 0$, then X_Γ ($= X$, by hypothesis) $= X_N$, by (5.8)(b), so $N\backslash X$ is finite by (5.6)(b). The last assertion, when Γ is discrete, is then clear.

If $\ell(N) = 0$ then, by (5.3), N has a non-empty tree Y of fixed points. Since $N \triangleleft \Gamma$, Y is Γ-invariant. By minimality, $Y = X$. This contradicts the assumption $N \neq \{1\}$.

For the next result we introduce some notation. Let $X' \subset X$ be a subtree. Define a retraction $\pi : VX \to VX'$ by:

$$\pi(x) = \text{the vertex of } X' \text{ nearest to } x (\in VX).$$

Then, for $x'\in VX'$,
$$\pi^{-1}(x') = VY(x'), \text{ where}$$

$$(Y(x'), x') \text{ is a rooted tree, attached to } X' \text{ at the root } x',$$

called the *normal tree* to X' in X at x'.

$$X = X' \cup \left(\bigcup_{x' \in VX'} Y(x') \right).$$

5.10   Proposition. *Suppose that $X' \subset X$ is a proper Γ-invariant subtree, and that Γ is an X-lattice. For each $x' \in VX'$, the normal tree $Y(x')$ to X' in X at x', is finite. Hence X has terminal vertices. Moreover the natural projection.*

$$p : X \to A = \Gamma \backslash X \supset A' = \Gamma \backslash X'$$

is injective on reduced paths from x' in $Y(x')$, and

$$A = A' \cup \left(\bigcup_{a' \in VA'} Y(a') \right)$$

where, if $a' = p(x')$, $x' \in VX'$ then $Y(a') = p(Y(x')) = \Gamma_{x'} \backslash Y(x')$, and the finite rooted tree $(Y(a'), a')$ is attached to A' at a'. Hence, if $\Gamma | X'$ is uniform (A' is finite), then Γ is uniform (A is finite).

Proof. Since X' is Γ-invariant, Γ preserves distance from X'. It follows that, if C is a reduced path from x' in $Y(x')$, then $p : X \to A$ is injective on C. Putting $a' = p(x')$ it follows that $Y(a') := p(Y(x'))$ is a tree attached to $A' = \Gamma \backslash X'$ at the root a', and clearly $Y(a')$ is independent of the choice of $x' \in p^{-1}(a')$.

It remains only to see that, when Γ is an X-lattice, the trees $Y(x')$ are finite. Since X is now (by assumption) locally finite, it suffices to show that each reduced path C in $Y(x')$ from x' is finite. The Γ-invariance of X' implies that the stabilizers along C descend starting from $\Gamma_{x'}$. Since p injects C into A, the finiteness of $\mathrm{Vol}(\Gamma \backslash\backslash X)$ implies that $p(C)$, hence also C, is finite, as claimed.

5.11   Proposition. *Let $\Gamma \le G = \mathrm{Aut}(X)$ be an X-lattice. If $\ell(G) \ne 0$, then $\ell(\Gamma) \ne 0$, and $X_\Gamma = X_G$.*

Proof. Replacing X by X_G, we may assume that G acts minimally on X. Since $\ell(G) \ne 0$, X is infinite. From (5.4) it follows that X has no terminal vertices. Now from (5.10) it follows that Γ acts minimally on X, and so, from (5.5), that $\ell(\Gamma) \ne 0$.

Remark. Proposition (5.11) is somewhat reminiscent of the Borel Density Theorem on lattices in algebraic groups.

5.12   Theorem. *Let X be an infinite locally finite tree which admits a lattice, and let $G = \mathrm{Aut}(X)$. The following conditions are equivalent:*
  (a) *Some X-lattice Γ acts minimally on X.*

(a′) *Every X-lattice Γ′ acts minimally on X.*

(b) *G acts minimally on X.*

(c) *X has no terminal vertices.*

Under these conditions, $\ell(\Gamma) \neq 0$ for all X-lattices Γ.

Proof. Let Γ be as in (a). Then $\ell(\Gamma) \neq 0$ by (5.5). Trivially, (a) $\Rightarrow$ (b). Assume (b). Then $\ell(G) \neq 0$. It then follows from (5.11) that every X-lattice Γ′ acts minimally, and $\ell(\Gamma′) \neq 0$. Whence (b) $\Rightarrow$ (a′). Clearly (a′) $\Rightarrow$ (a); and (b) $\Rightarrow$ (c) by (5.4). Finally (c) $\Rightarrow$ (a′) by (5.10).

5.14 Abelian actions

Let $\varphi : \Gamma \to \mathbf{Z}$ be a homomorphism. Then Γ acts on the linear tree $X(\varphi) = \mathbf{Z}$ by translation: $gn = n + \varphi(g)$ for $g \in \Gamma$. Clearly then

$$\ell_{X(\varphi)}(g) = |\varphi(g)|.$$

Call a Γ-tree X *abelian* if $\ell_X = |\varphi|$ for some homomorphism φ as above; φ is unique up to a factor ± 1 ([AB], (1.4)). Moreover there is a Γ-equivariant morphism $X \to X(\varphi)$, unique up to a translation of $X(\varphi)$ ([AB], p. 344).

For a Γ-tree X without inversions, the following conditions are equivalent (cf. [AB], §7 or [CM]):

(a) *X is abelian.*

(b) $\ell(ghg^{-1}h^{-1}) = 0$ for all $g, h \in \Gamma$ ($\ell = \ell_X$).

(c) $\ell(gh) \leq \ell(g) + \ell(h)$ for all $g, h \in \Gamma$.

(d) $X_g \cap X_h \neq \emptyset$ for all $g, h \in \Gamma$.

(e) Γ fixes a vertex or an end of X.

5.15 Non-abelian actions (cf. [AB], (7.13), [CM], [BJ], (1.7))

Theorem. *Let X, Y be minimal non-abelian Γ-trees without inversions. If $\ell_X = \ell_Y$, then there is a unique Γ-equivariant morphism $\varphi : X \to Y$, and it is an isomorphism. In particular, $\mathrm{Aut}_\Gamma(X) = \{1\}$.*

5.16 Abelian discrete actions

Let $\Gamma \leq G = \mathrm{Aut}(X)$ be a *discrete* subgroup with abelian action, say $\ell(g) = |\tau(g)|$ ($g \in \Gamma$), where $\tau : \Gamma \to \mathbf{Z}$ is a homomorphism. Then Γ is without inversion, since an inversion fixes no vertex or end (cf. (5.13)(e)). Clearly:

(1)   If Γ fixes a vertex then Γ is finite. If, further, Γ is an X-lattice, then X is finite.

Suppose now that:

(2) Γ fixes no vertex of X.

Then:

(3)   If $\ell(\Gamma) = 0$ then the action of Γ on X is "parabolic," i.e., Γ fixes a unique
        end, ε ((4.10)(1)). Moreover all horoballs centered at ε are Γ-invariant (cf.
        (9.2) below). If, further, Γ is an X-lattice, then X is a "parabolic tree," i.e., ε
        is the unique end of X.

The first part follows from (5.2). The last assertion follows from Theorem (9.7)
below.

(4)   If $\ell(\Gamma) \neq 0$, then X_Γ (cf. (5.11)) is a linear tree on which Γ acts by translation,
        with $\Gamma^0 (:= Ker(\tau))$ a finite group acting trivially on X_Γ; Γ fixes both ends
        of X_Γ.

This follows from (9.4)(4) below. In conclusion, we observe:

5.16   **Proposition.** *Let X be a locally finite tree and Γ a non-uniform X-lattice.*
       (a) *Γ has arbitrarily large finite subgroups, hence Γ is not virtually torsion free.*
       (b) *Γ is not finitely generated.*

Since $\mathrm{Vol}(\Gamma \backslash\backslash X) = \sum_{x \in \Gamma \backslash VX} 1/|\Gamma_x|$ is finite, and $\Gamma \backslash X$ is infinite, the groups Γ_x
have unbounded size, whence (a).
       Suppose that Γ is finitely generated. If $\ell(\Gamma) = 0$ then, by (5,3), Γ is finite,
contradicting (a). Thus $\ell(\Gamma) \neq 0$. It follows then from (5.6) that $\Gamma \backslash X_\Gamma$ is finite.
       According to (5.10), $\Gamma \backslash X$ is obtained from $\Gamma \backslash X_\Gamma$ by attaching a finite rooted tree
to each vertex of $\Gamma \backslash X_\Gamma$. It follows that $\Gamma \backslash X$ is finite, contradicting the assumption
that Γ is non-uniform.

5.17   **Corollary.** *Let H be a simple rank 1 Lie group over a non-archimedean
local field. Let Γ be a non-uniform H-lattice. Then Γ has unbounded torsion and is
not finitely generated.*

This follows by applying (5.16) to the action of H on its Bruhat–Tits tree X, with
$H \backslash X$ an edge. It was proved by Behr ([Be]) for arithmetic lattices, and in general by
Lubotzky ([L3]).

Chapter 6

Centralizers, Normalizers, and Commensurators

6.0 Introduction

Consider a group G and a subgroup $\Gamma \leq G$. We are interested here in the centralizer $Z_G(\Gamma)$, normalizer $N_G(\Gamma)$, and commensurator $C_G(\Gamma)$ (see (6.1)). In (6.2)–(6.4), G will be a topological group. From (6.5) on, $G = \mathrm{Aut}(X)$, where X is a tree. Whenever we assume that Γ is an X-lattice, it is implicity assumed also that X is locally finite.

When $\Gamma \leq G = \mathrm{Aut}(X)$ is an X lattice, our results include the following:

(1) $Z_G(\Gamma)$ is a closed subgroup of G. If Γ acts minimally an X then $Z_G(\Gamma) = \{1\}$ unless $X = \circ\!\!-\!\!\circ$ or X is a linear tree on which Γ-acts by translation ((6.5)). If the Γ action on X is non-abelian then $Z_G(\Gamma)$ is a product of finite groups and is finite if Γ is uniform ((6.7)).

(2) $N_G(\Gamma)$ is a closed subgroup of G, and $N_G(\Gamma)/\Gamma$ is a profinite group, finite if Γ is uniform ((6.8)).

(3)   If $C_G(\Gamma)$ is dense in G then Γ is residually finite, except perhaps in the following situation:

$$\Gamma \text{ is non-uniform,} \quad \ell(\Gamma) \neq 0,$$
$$\Gamma | X_\Gamma \text{ is residually finite, and}$$
$$Ker(\Gamma \xrightarrow{res} \Gamma | X_\Gamma) \text{ is finite.}$$

(See (6.18).)

The proof of (3) makes use of the Tits Simplicity Theorem ((6.12)).

6.1 Notation

Let G be a group. For $g \in G$ we define $ad(g)$ $(= ad_G(g) \in \mathrm{Aut}(G))$ by $ad(g)(x) = gxg^{-1}$. We have the exact sequence

$$1 \to Z(G) \to G \xrightarrow{ad_G} \mathrm{Aut}(G) \to \mathrm{Out}(G) \to 1.$$

We call G *inn-exact* if $ad_G : G \to \mathrm{Aut}(G)$ is an isomorphism. (In the literature one sometimes finds the term "complete" for this notion.)

Let $\Gamma \leq G$. We define

(1) $\mathrm{Aut}(G; \Gamma) = \{\alpha \in \mathrm{Aut}(G) |\ \alpha \Gamma = \Gamma\}.$

More generally, if $(\Gamma_n)_n$ is a family of subgroups we put

(2) $\mathrm{Aut}(G; (\Gamma_n)_n) = \bigcap_n \mathrm{Aut}(G; \Gamma_n).$

The *centralizer* and *normalizer* are denoted

(3)
$$Z_G(\Gamma) = \{g \in G |\ gh = hg \quad \forall\, h \in \Gamma\},\ \text{resp.}$$
$$N_G(\Gamma) = \{g \in G |\ g \Gamma g^{-1} = \Gamma\}.$$

For $\Gamma' \leq G$ we put

(4) $Z_{\Gamma'}(\Gamma) = \Gamma' \cap Z_G(\Gamma),\ N_{\Gamma'}(\Gamma) = \Gamma' \cap N_G(\Gamma).$

For $g \in N_G(\Gamma)$ put $ad_\Gamma(g) = (ad_G(g)|\Gamma) \in \mathrm{Aut}(\Gamma)$, whence an exact sequence

(5) $1 \to Z_{\Gamma'}(\Gamma) \to N_{\Gamma'}(\Gamma) \xrightarrow{ad_\Gamma} \mathrm{Aut}(\Gamma).$

We write

(6) $\Gamma \leq_{virt} \Gamma'$ if $\Gamma \cap \Gamma'$ has finite index in Γ,   and

(7) $\Gamma =_{virt} \Gamma'$ if $\Gamma \leq_{virt} \Gamma'$ and $\Gamma' \leq_{virt} \Gamma$,

in which case Γ and Γ' are said to be *commensurable* (or "*virtually equal*"). The *G-commensurator* of Γ is

(8) $C_G(\Gamma) = \{g \in G |\ g \Gamma g^{-1} =_{virt} \Gamma\}.$

We further introduce

(9)
$$K_G(\Gamma) := \bigcap_{g \in C_G(\Gamma)} g \Gamma g^{-1}$$

$\qquad\qquad$ = the largest normal subgroup of $C_G(\Gamma)$ contained in Γ,   and

$$H_G(\Gamma) := \left\langle \bigcup_{g \in C_G(\Gamma)} g\Gamma g^{-1} \right\rangle$$

(10)

= the smallest normal subgroup of $C_G(\Gamma)$ containing Γ .

Clearly

$$K_G(\Gamma) \leq \Gamma \leq H_G(\Gamma) \leq C_G(\Gamma),$$

(11) $$K_G(\Gamma), H_G(\Gamma) \triangleleft C_G(\Gamma), \text{ and}$$

$$Z_G(\Gamma) \cdot \Gamma \triangleleft N_G(\Gamma) \leq C_G(\Gamma).$$

Consider also

$$P(\Gamma) := Ker(\Gamma \to \hat{\Gamma}), \text{ where}$$

(12) $$\hat{\Gamma} = \text{the profinite completion of } \Gamma$$

= the inverse limit of the finite quotients of Γ.

Evidently

$$P(\Gamma) \leq K_G(\Gamma), \text{ so, if } K_G(\Gamma) = \{1\}$$

(13)

then Γ is residually finite.

If $\Gamma =_{virt} \Gamma'$ then clearly $P(\Gamma) = P(\Gamma')$. If $g \in C_G(\Gamma)$ then $gP(\Gamma)g^{-1} = P(g\Gamma g^{-1}) = P(\Gamma)$, whence:

(14)   (a) $P(\Gamma) \triangleleft C_G(\Gamma)$.
          (b) If $\hat{\Gamma}$ is finite then $\widehat{P(\Gamma)} = \{1\}$, and $C_G(\Gamma) = N_G(P(\Gamma)) = C_G(P(\Gamma))$.

6.2   Proposition. *Let G be a topological group, and $\Gamma \leq G$.*
  (a) *$Z_G(\Gamma)$ is a closed subgroup of G.*
  (b) *If Γ is closed, then so also is $N_G(\Gamma)$.*
  (c) *If Γ is discrete, then $Z_{N_G(\Gamma)}(\Gamma')$ is open in $N_G(\Gamma)$ for every finitely generated $\Gamma' \leq \Gamma$. Hence, if $Z_G(\Gamma')$ is discrete for some finitely generated $\Gamma' \leq \Gamma$, then $N_G(\Gamma)$ is discrete.*
  Suppose that Γ is closed. Then:
  (d) *$K_G(\Gamma)$ is a closed normal subgroup of $\overline{C_G(\Gamma)}$ (the closure of $C_G(\Gamma)$). If $K_G(\Gamma)$ is discrete (e.g., if Γ is discrete) then $Z_{\overline{C_G(\Gamma)}}(\Gamma')$ is open in $\overline{C_G(\Gamma)}$ for all finitely generated $\Gamma' \leq K_G(\Gamma)$.*

Proof. For $g \in G$ define $c_g(h) = hgh^{-1}, c_g : G \to G$. Since c_g is continuous, $Z_G(g) = c_g^{-1}(g)$ is closed, hence also $Z_G(\Gamma) = \bigcap_{g \in \Gamma} Z_G(g)$.
    If Γ is closed so also is

$$N' = \bigcap_{g \in \Gamma} c_g^{-1}(\Gamma) = \{h \in G \mid h\Gamma h^{-1} \leq \Gamma\}$$

as well as $N_G(\Gamma) = N' \cap N'^{-1}$.

It suffices to prove (c) when $\Gamma' = \langle g \rangle$, a cyclic group, $g \in \Gamma$. Then $c_g(N_G(\Gamma)) \subset \Gamma$, which is assumed discrete. Let $n \in N = N_G(\Gamma)$ centralize g. Let U be a neighborhood of g in G such that $U \cap \Gamma = \{g\}$. Since c_g is continuous, we can choose a neighborhood V of n in N such that $c_g(V) \subset U$. But $c_g(V) \subset c_g(N) \subset \Gamma$, so $c_g(V) \subset U \cap \Gamma = \{g\}$. Hence V centralizes g, thus showing that $Z_N(g)$ is open in N.

If Γ is closed it is immediate from (6.1)(9) that $K_G(\Gamma)$ is closed, hence so also is $N_G(K_G(\Gamma))$, by (b). The latter contains $C_G(\Gamma)$, hence also $\overline{C_G(\Gamma)}$. Now the last part of (d) follows from (c), applied to $K_G(\Gamma)$ in place of Γ.

6.3   **Corollary.** *Let G be a Lie group and $\Gamma \le G$. If Γ and $Z_G(\Gamma)$ are discrete, so also is $N_G(\Gamma)$*

Proof. Since $Z_G(\Gamma)$ is discrete, it has Lie algebra $\mathrm{Lie}(Z_G(\Gamma)) = 0$. Then, by finite dimensionally of $\mathrm{Lie}(G)$, $\mathrm{Lie}(Z_G(\Gamma')) = 0$ for some finitely generated $\Gamma' \le \Gamma$, so we can apply (5.2)(c).

6.4   **Corollary.** *Let G be a topological group and $\Gamma \le G$ a discrete subgroup, satisfying*

(1) *G contains no non-trivial discrete normal subgroups, and*

(2) *$C_G(\Gamma)$ is dense in G.*

Then $K_G(\Gamma) = \{1\}$, and so Γ is residually finite.

In fact, by (6.2)(d) and (2) $K_G(\Gamma)$ is a discrete normal subgroup of G, hence trivial, by (1), so (6.1)(13) applies.

Remark. If, in (6.4) we relax (1) to:

(1′)   Discrete normal subgroups of G are finite,

then the conclusion of (6.4) changes to:

$$K_G(\Gamma) \text{ is finite, hence so also is } P(\Gamma) = Ker(\Gamma \to \hat{\Gamma}).$$

Henceforth, we assume that

(Hyp) $\begin{cases} G = \mathrm{Aut}(X), \text{ where} \\ X = \text{a tree, and} \\ \Gamma \le G. \end{cases}$

The topology on G declares $g, g' \in G$ to be close if g and g' agree on a large finite subtree.

6.5   **Proposition** (cf. [BJ], (1.5)). *If Γ acts minimally on X then $Z_G(\Gamma)$ ($= \mathrm{Aut}_\Gamma(X)$* *(cf. (5.1)(8))) $= \{1\}$, except in the following cases.*

(i) $X = \circ\!\!-\!\!-\!\!\circ$ and $Z_G(\Gamma) = G$.

(ii) X is a linear tree on which Γ acts by translations, and $Z_G(\Gamma)$ is the full group of translations.

Proof. Let $z \in Z_G(\Gamma)$, $z \neq 1$. Then X_z is Γ-invariant ((5.1)(7)), so $X_z = X$, by minimality. If z fixes a vertex, then $z = 1$, contrary to assumption. If z is an inversion we have case (i). If z is hyperbolic we have case (ii), since the centralizer of a non-trivial translation in the dihedral group G is the full group of translations.

6.6   Non-minimal centralizers

Let $X' \subset X$ be a subtree. Define $\pi : VX \to VX'$ by $\pi(x) =$ the vertex of X' nearest to x. Then, for $x \in VX'$, $\pi^{-1}(x)$ is the vertex set of a tree that we denote $Y(x)$ (or $Y_{X'}(x)$); then $(Y(x), x)$ is a rooted tree.

Let

(1) $$G_{X'} = Fix_G(X') := \{g \in G \mid gx = x \ \forall \ x \in VX'\}.$$

For $x \in VX'$, we have a restriction homomorphism

(2) $$\partial_x : G_{X'} \to \text{Aut}(Y(x), x), \quad \text{denoted}$$
                     $$\partial_x(g) = g(x) := g \mid Y(x).$$

These assemble to give

(3) $$\partial_{X'} : G_{X'} \xrightarrow{(\partial_x)_x} \prod_{x \in VX'} \text{Aut}(Y(x), x),$$

which is clearly an isomorphism. Let $\Gamma \leq G$. We put

(4) $$Fix_\Gamma(X') = \Gamma \cap G_{X'}, \quad \text{and}$$
                     $$\Gamma(x) \ (\text{or} \ \Gamma_{X'}(x)) = \partial_x(Fix_\Gamma(X')) \ \text{for} \ x \in VX'. \ (\text{See (2).})$$

For example, we have

(5) $$G_{X'}(x) = \text{Aut}(Y(x), x).$$

Suppose now that

(6) $$X' \text{ is } \Gamma\text{-invariant.}$$

Then we have the exact sequence

$$1 \to Fix_\Gamma(X') \to \Gamma \to \Gamma \mid X' \to 1$$
$$(\leq \text{Aut}(X'))$$

Let

(7) $Z = Z_G(\Gamma)$,   and assume that X' is also Z-invariant.

Then

(8) $Z|X' \leq Z_{\mathrm{Aut}(X')}(\Gamma|X')$

If Γ acts minimally on X' then the latter group is trivial, but for the exceptional cases noted in (6.5).

For $x \in VX'$ and $g \in \Gamma$, we have the isomorphism

$$g : Y(x) \xrightarrow{\cong} Y(gx).$$

For $z \in Fix_Z(X')$ we have a commutative diagram

$$
\begin{array}{ccc}
Y(x) & \xrightarrow{g} & Y(gx) \\
z \uparrow & & \uparrow z \\
Y(x) & \xrightarrow{g} & Y(gx)
\end{array}
$$

whence $z(gx) = gz(x)g^{-1} = ad(g)(z(x))$. Thus one sees that

(9) $Z(x) = \mathrm{Aut}_{\Gamma_x}(Y(x), x) \leq \mathrm{Aut}(Y(x), x) = G_{X'}(x),$

and, for $g \in \Gamma$,

(10) $ad(g) : Z(x) \xrightarrow{\cong} Z(g(x))$ is an isomorphism that depends only on g mod Γ_x.

Writing thus

(11) $G_{X'} = \prod_{x \in VX'} G_{X'}(x) \cong \prod_{x \in \Gamma \backslash VX'} G_{X'}(x)^{\Gamma \cdot x},$

the conjugation action of Γ permutes factors of $G_{X'}(x)^{\Gamma \cdot x}$ via its action on the Γ-set $\Gamma \cdot x$. From this and (9),(10) we conclude that

(12) $Fix_Z(X') \cong \prod_{x \in \Gamma \backslash VX'} \mathrm{Aut}_{\Gamma_x}(Y(x), x).$

Observe next that,

(13)   If Γ is an X-lattice, then the tree $Y(x)$ is finite for all $x \in VX'$, and hence $\mathrm{Aut}(Y(x), x)$ is a finite group.

In fact, let γ be a reduced path in $Y(x)$ starting at x. Since X' is Γ-invariant, the Γ stabilizers, starting with Γ_x, decrease along γ. Moreover, since Γ preserves distance from X', γ injects into $\Gamma \backslash X$. Since $\mathrm{Vol}(\Gamma \backslash\backslash X)$ is finite, by assumption, the path γ must be finite. Thus $(Y(x), x)$ is a locally finite rooted tree with all paths from the root finite, hence $Y(x)$ is finite (Konig's Theorem).

6.7   Proposition. *Suppose that the action of Γ on X is non-abelian. Put $X' = X_\Gamma$, (cf. (5.7)), and adopt the notation of (6.6). Then*

(1)
$$Z_G(\Gamma) \cong \prod_{x \in \Gamma \backslash VX_\Gamma} \mathrm{Aut}_{\Gamma_x}(Y(x), x).$$

If Γ is an X-lattice, then each of the trees $Y(x)$, and groups $\mathrm{Aut}_{\Gamma_x}(Y(x), x)$, is finite $(x \in VX_\Gamma)$; hence $Z_G(\Gamma)$ is a product of finite groups. If Γ is a uniform lattice, then $\Gamma \backslash X_\Gamma$ is finite, so $Z_G(\Gamma)$ is a finite group.

These assertions follow from (6.6) once we observe that X_Γ is $Z_G(\Gamma)$-invariant ((5.7)), and $Z_G(\Gamma)|X_\Gamma = Id$ because $Z_{\mathrm{Aut}(X_\Gamma)}(\Gamma|X_\Gamma) = \{1\}$, since the action is minimal and non-abelian ((6.5)).

6.8   Proposition. *Let Γ be an X-lattice and $N = N_G(\Gamma)$.*
   (a) *For $x \in VX$, $\Gamma \backslash N / N_x \; (= N / \Gamma \cdot N_x)$ is finite.*
   (b) *N / Γ is a profinite group.*
   (c) *If Γ is uniform then N / Γ is finite.*

Proof. N acts on $A = \Gamma \backslash X$ so that the projection $p : X \to A$ is N-equivariant. Define
$$m : VA \to \mathbf{R}, \quad m(p(x)) = 1/|\Gamma_x|.$$
Since
$$\mathrm{Vol}(\Gamma \backslash\backslash X) \; (= \sum_{a \in VA} m(a)) < \infty$$
it follows that m has finite fibers. Clearly m is an N-invariant function. Hence N has finite orbits in A. Thus, for $x \in VX$ and $a = p(x)$, we have the finiteness of
$$N \cdot a = \Gamma \backslash (N \cdot x) \cong \Gamma \backslash N / N_x,$$
whence (a). Since N is a closed subgroup of G ((5.2)(b)), N_x is an open profinite subgroup of N. Then $N_x \cdot \Gamma / \Gamma (\cong N_x / \Gamma_x)$, is an open profinite subgroup of finite index in N / Γ, and so N / Γ is profinite, whence (b).

Finally, assertion (c) follows from [BK], (6.4).

Remarks. 1. When Γ is a non-uniform X-lattice we shall see examples (in Chapters 9–10) when $N_G(\Gamma)/\Gamma$ can be either finite (even trivial) or infinite.

2. An action of a group Γ on a *set* V is called *strict* if, for each $x \in V$, Γ_x fixes only x. An action of Γ on a *tree* X is called *strict* if Γ acts strictly on VX and, for all

$x \in VX$, Γ_x acts strictly on $E_0(x)$. When X is locally finite and $\Gamma \leq G = \mathrm{Aut}(X)$ is discrete and acts strictly on X, it then follows from [BL], Theorem (4.4), that

$$(1) \qquad\qquad ad_\Gamma : N_G(\Gamma) \xrightarrow{\cong} \mathrm{Aut}(\Gamma)$$

is an isomorphism. An example of this is an amalgam $\Gamma = \Gamma_0 *_{\Gamma_e} \Gamma_1$, the fundamental group of a graph of finite groups $\overset{\Gamma_0}{\circ}\!\!\rule[0.5ex]{2.5em}{0.4pt}\!\!\overset{\Gamma_1}{\circ}$ $\overset{\Gamma_e}{}$. The action of Γ on the covering tree X will be strict iff $\Gamma_e = N_{\Gamma_i}(\Gamma_e) \neq \Gamma_i (i = 0, 1)$ (cf. [BL], (3.8)). In this case we conclude from (1) and (6.8)(c) that $N_G(\Gamma)/\Gamma \cong \mathrm{Out}(\Gamma)$ is a finite group, even though Γ is finitely generated and virtually free of rank > 1.

6.9 N/Γ, for minimal non-abelian actions

With $\Gamma \leq G = \mathrm{Aut}(X)$ and $N = N_G(\Gamma)$, *assume now that the action of Γ on X is minimal and non-abelian* ((5.14)) (*and without inversions*). Let $\ell : \Gamma \to \mathbf{Z}$ be the hyperbolic length function ((5.1)(8)) and

$$(1) \qquad\qquad \mathbf{A} = (A, \mathcal{A}) = \Gamma \backslash\backslash X$$

a quotient graph of groups (cf. (2.4)). Then, from [BJ] we obtain the following more precise information about N and about

$$(2) \qquad\qquad W = N/\Gamma.$$

First ([BJ], (1.11))

$$(3) \qquad\qquad N = \mathrm{Aut}(\Gamma)_\ell := \{\varphi \in \mathrm{Aut}(\Gamma) | \ell \circ \varphi = \ell\}, \quad \text{so}$$

$$(4) \qquad\qquad W = \mathrm{Out}(\Gamma)_\ell.$$

Further ([BJ], §8), W admits a filtration,

$$(5) \qquad\qquad W \rhd W^A \rhd W^{(V)} \rhd W^{(V,E)} \rhd \{1\},$$

whose quotients admit the following descriptions.

$$(6) \qquad \begin{array}{l} W^A = N_{G_\Gamma}(\Gamma)/\Gamma, \quad (G_\Gamma \text{ as in (3.3)), and} \\ W/W^A \leq \mathrm{Aut}(A). \end{array}$$

(When A is finite, e.g., if Γ is a uniform X-lattice, $\mathrm{Aut}(A)$ is finite.)

$$(7) \quad \begin{array}{l} W^A/W^{(V)} \leq \displaystyle\prod_{a \in VA} \mathrm{Out}^E(\mathcal{A}_a), \quad \text{where} \\[1em] \mathrm{Out}^E(\mathcal{A}_a) = \mathrm{Aut}^E(\mathcal{A}_a)/ad(\mathcal{A}_a), \quad \text{and} \\[0.5em] \mathrm{Aut}^E(\mathcal{A}_a) = \{\varphi \in \mathrm{Aut}(\mathcal{A}_a) | \; \varphi\alpha_e \mathcal{A}_e \text{ is } \mathcal{A}_a\text{-conjugate to } \alpha_e \mathcal{A}_e \; \forall \, e \in E_0(a)\}. \end{array}$$

(See [BJ], (6.7) for a precise description of $W^A/W^{(V)}$ inside the above product.)

$$(8) \qquad W^{(V)}/W^{(V,E)} \cong \prod_{\{e,\bar{e}\}} \frac{N(e) \cap N(\bar{e})}{ad(\mathcal{A}_e)}.$$

Here $\{e, \bar{e}\}$ varies over the geometric edges of A. Say $\partial_0 e = a$. Then $\alpha_e : \mathcal{A}_e \to \mathcal{A}_a$ is injective, so $ad_{\alpha_e \mathcal{A}_e}(N_{\mathcal{A}_a}(\alpha_e \mathcal{A}_e)) \leq \mathrm{Aut}(\alpha_e \mathcal{A}_e)$ defines a subgroup $N(e) \leq \mathrm{Aut}(\mathcal{A}_e)$, containing $ad(\mathcal{A}_e)$. We similarly have $N(\bar{e}) \leq \mathrm{Aut}(\mathcal{A}_{\bar{e}}) = \mathrm{Aut}(\mathcal{A}_e)$.

$$W^{(V,E)} = \prod_{a \in VA} Z(a), \quad \text{where}$$

(9)

$$Z(a) = Coker(Z(\mathcal{A}_a) \xrightarrow{\Delta_a} \prod_{e \in E_0(a)} \frac{Z_{\mathcal{A}_a}(\alpha_e \mathcal{A}_e)}{\alpha_e(Z(\mathcal{A}_e))}),$$

and Δ_a is induced by the inclusion $Z(\mathcal{A}_a) \leq Z_{\mathcal{A}_a}(\alpha_e \mathcal{A}_e))$ $(e \in E_0(a))$.

6.10   Some normal subgroups

We review here, and in the following two sections, some important results of Tits ([Ti1]).

Call an action of a group L on a set E *saturated* if $Im(L \to \mathrm{Aut}(E))$ is the complete direct product of the full symmetric groups on each of the L-orbits.

Let $H \leq G = \mathrm{Aut}(X)$. We call the action of H in X *locally saturated* if it satisfies

(LS)   For all $x \in VX$, the action of H_x on $E_0(x)$ is saturated.

For example G_H (cf. (3.3)) clearly satisfies (LS). We introduce the normal subgroup of H,

$$(1) \qquad H^- = \left\langle \bigcup_{e \in EX} H_e \right\rangle$$

and

$$(2) \qquad H^Y = \left\langle \bigcup_{x \in VX, \deg(x) \geq 3} H_x \right\rangle.$$

The superscripts "$-$" and "Y" are meant to suggest an edge, and a vertex of degree ≥ 3, respectively.

Proposition.

(a) $H^- \leq H^Y$.

(b) *If the action of H on X satisfies (LS) then $H^Y = H^-$.*

(c) H/H^Y *is a free product of cyclic groups, each of order ∞ or 2.*

Proof of (a). Let $e \in EX$, and let X' denote the fixed tree of H_e. If $\deg_X(x) \geq 3$ for some $x \in VX'$, then $H_e \leq H_x \leq H^Y$. So suppose instead that $\deg_X(x) \leq 2$ for all $x \in VX'$. If $\deg_X(x) = 2$, then H_e, since it fixes one of the two edges at x, fixes both of them, so $\deg_{X'}(x) = 2$. Thus $\deg_X(x) = \deg_{X'}(x)$ for all $x \in VX'$. It follows that X' is a connected component of X, hence $X' = X$, and $H_e = \{1\} \leq H^Y$.

Proof of (b). Let $x \in VX$. Then $H_x | E_0(x)$, is, by (LS), isomorphic to $\Pi_e \, S(H_x \cdot e)$, where $S(E)$ denotes the symmetric group on a set E, and e varies over $H_x \backslash E_0(x)$. We have $\deg(x) = \sum_e |H_x \cdot e|$. If $\deg(x) \geq 3$, the above group is generated by elements with fixed points, whence H_x is generated by elements fixing some edge in $E_0(x)$, so $H_x \leq H^-$.

Proof of (c). Let X' be the barycentric subdivision of X, on which H acts without inversions. Then $L := H/H^Y$ acts on $T = H^Y \backslash X'$. Since H^Y is generated by elements with fixed points, T is still a tree (cf. [B3], (1.19), Ex. (3)). Let $p : X' \to T$ be the projection, and let $x \in VX'$. If $\deg_{X'}(x) \geq 3$, then $H_x \leq H^Y$, so $L_{p(x)} = \{1\}$. If $\deg_{X'}(x) = 2$, then $K = Ker(H_x \to Aut(E_0(x)))$ has index ≤ 2 in H_x, and $K \leq H^Y$, by (a). Thus $L_{p(x)}$ has order ≤ 2. For $e \in EX'$, since $H_e \leq H^- \leq H^Y$, $L_{p(e)} = \{1\}$. It follows that $L \backslash\backslash T$ is a graph of groups with trivial edge groups, and vertex groups of orders ≤ 2. Hence (cf. [B3], (1.19), Example (2)), $L \cong \pi_1(L \backslash\backslash T, *)$ has the structure indicated in (c).

6.11   The Tits Independence Condition

Let $H \leq G = Aut(X)$, as above. Let X' be a subtree. As in (6.6)(3),(4) we have groups

(1) $$H_{X'}(x) = Fix_H(X')|Y(x) \quad (x \in VX')$$

and the injective homomorphism

(2) $$\partial_{X'} : Fix_H(X') \to \prod_{x \in VX'} H_{X'}(x).$$

We say that "*H is independent along X'*" if,

$(I)_{X'}$ The homomorphism $\partial_{X'}$ in (2) is an isomorphism.

For example the group G_H (see (3.3)) clearly satisfies $(I)_{X'}$ for all $X' \subset X$. We write $(I)_e$ $(e \in EX)$ for $(I)_{X'}$ when $EX' = \{e, \bar{e}\}$.

Proposition. *Suppose that H satisfies* $(I)_e$ *for all* $e \in EX$. *Then:*

(a) *H satisfies* $(I)_{X'}$, *for all finite trees* $X' \subset X$.

(b) *If further H satisfies condition (LS) of (6.10) then H is dense in* G_H.

Proof of (a). We argue by induction on $n = |VX'|$. The case $n = 1$ is vacuously true, and the case $n = 2$ is just $(I)_e$. So suppose that we have $(I)_{X'}$, with $n > 1$, and we want to establish $(I)_{X''}$, where X'' is obtained from X' by attaching a new edge e at $\partial_0 e = x \in VX'$; put $y = \partial_1 e$.

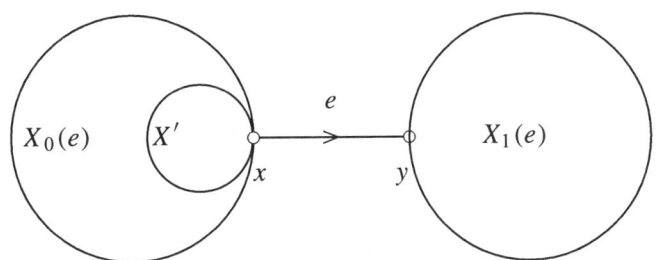

Let $g \in F'' := Fix_H(X'')$. It suffices to show the following:

$(*)_z$ Given $z \in VX''$, there is an $h \in F''$ such that $h = g$ on $Y_{X''}(z)$ and $h = Id$ on $Y_{X''}(w)$ for $w \in VX'' - \{z\}$.

Note that

(i) $Y_{X''}(w) = Y_{X'}(w)$ for $w \in VX' - \{x\}$,

and

(ii) $Y_{X'}(x) = Y_{X''}(x) \cup \{e, \bar{e}\} \cup Y_{X''}(y)$.

Put $F' := Fix_H(X')$ $(\geq F'')$. To prove $(*)_z$ we distinguish three cases.

Case $z \in VX' - \{x\}$. By induction there is an $h \in F'$ such that $h = g$ on $Y_{X'}(z) = Y_{X''}(z)$ (by (i)), and $h = Id$ on $Y_{X'}(w)$ for $w \in VX' - \{z\}$. For $w \neq x$, $Y_{X'}(w) = Y_{X''}(w)$. Consider $w = x$: Since $g \in F''$, $ge = e$, hence $he = e$, i.e., $h \in F''$. Further $h = Id$ on $Y_{X'}(x)$, so by (ii), $h = Id$ on $Y_{X''}(x)$ and on $Y_{X''}(y)$.

Case $z = x$. Again choosing $h \in F'$ as in the previous case, we have $h \in F''$, $h = Id$ on $Y_{X'}(w) = Y_{X''}(w)$ for $w \in VX' - \{x\}$, and $h = g$ on $Y_{X'}(x) = Y_{X''}(x) \cup \{e, \bar{e}\} \cup X_1(e)$. Now by $(I)_e$ we can choose $h_1 \in H_e$ so that $h_1 = h$ on $X_0(e)$ and $h_1 = Id$ on $X_1(e)$. It follows then that $h_1 \in F''$, $h_1 = g$ on $Y_{X''}(x)$, and $h_1 = Id$ on $Y_{X''}(w)$ for $w \in VX'' - \{x\}$.

Case $z = y$. By $(I)_e$ we can choose $h \in H_e$ so that $h = g$, on $X_1(e) = Y_{X''}(y)$, and $h = Id$ on $X_0(e)$, hence on X'', so $h \in F''$. Further, $h = Id$ on $Y_{X''}(w)$ $(\subset X_0(e))$ for $w \in VX'' - \{y\}$.

Proof of (b). Given a finite tree $X' \subset X$, and $g \epsilon G_H$, we must find $h \epsilon H$ so that $h = g$ on X'. We argue by induction on $n = |VX'|$. The cases $n = 1, 2$, follow directly from the definition of G_H:

$$G_H = \{g \epsilon G \mid p \circ g = g\}, \quad \text{where} \quad p : X \to H \backslash X.$$

So suppose that $n > 2$. Let $e_+, e_- \epsilon E X'$ be distinct edges so that $\partial_1(e_\pm)$ are terminal vertices of X'.

Case 1. For all choices of $e_\pm$ as above we have $\partial_0 e_+ = \partial_0 e_-$. In this case $X' \subset B_{x_0}(1)$, the ball of radius 1 centered at some $x_0 \epsilon VX$. Then $EX' \subset E_0(x_0) \cup \overline{E_0(x_0)}$. Condition (LS) and the definition of G_H assures us that $H_x|E_0(x_0) = (G_H)_x|E_0(x_0)$. Now first choose $h_0 \epsilon H$ so that $h_0 x_0 = g x_0$. Put $g_1 = h_0^{-1} g \epsilon (G_H)_{x_0}$. As just observed, there is an $h_1 \epsilon H_1$ agreeing with g_1, on $E_0(x_0)$, hence $h = h_0 h_1 \epsilon H$ agrees with $h_0 g_1 = g$ on $E_0(x_0)$, hence also on X'.

Case 2. We can choose $e_\pm$ above so that $\partial_0 e_+ = x_+ \neq x_- = \partial_0 e_-$. Then we have the schematic picture of X':

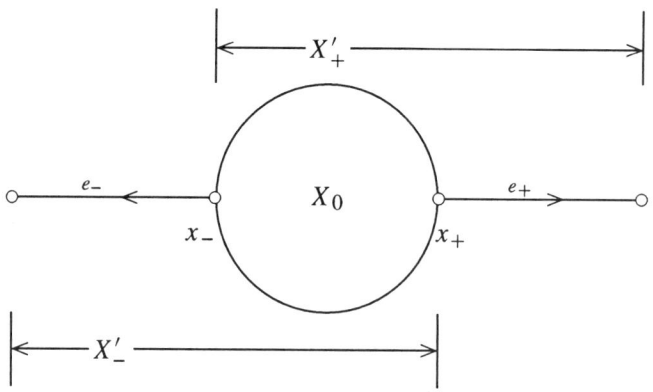

Put $X'_+ = X' - \{e_-, \bar{e}_-, \partial_1 e_-\}$, and $X'_- = X' - \{e_+, \bar{e}_+, \partial_1 e_+\}$. By induction can find $h_0 \epsilon H$ so that $h_0 = g$ on X_0. Put $g_1 = h_0^{-1} g$; then $g_1 = Id$ on X_0. By induction again we can find $h_\pm \epsilon H$ so that $h_+ = g_1$ on X_+, and $h_- = g_1$ on X_-. Then $h_+, h_- \epsilon F = Fix_H(X_0)$. By part (a), we can find $h_1 \epsilon F$ so that $h_1 = h_+$ on $Y_{X_0}(x_+)$ and $h_1 = h_-$ on $Y_{X_0}(x_-)$. Hence $h_1 e_+ = h_+ e_+ = g_1 e_+$ and $h_1 e_- = h_- e_- = g_1 e_-$, so $h_1 = g_1$ on X'. Therefore $h = h_0 h_1 \ (\epsilon H)$ agrees with $h_0 g_1 = g$ on X'.

6.12   **Tits Simplicity Theorem** ([Ti1]), Théorème 4.5). *Let X be a tree and $H \leq G = \mathrm{Aut}(X)$. Suppose that*
  ○ *H acts minimally on X,*
  ○ *H fixes no end of X, and*

o H satisfies condition $(I)_{X'}$ (of (6.11)) for all linear subtrees $X' \subset X$.

Let $K \leq H$ be a subgroup normalized by $H^- = \langle \bigcup_{e \in EX} H_e \rangle$ (cf. (6.10)(1)). Then either $K = \{1\}$ or $H^- \leq K$. In particular, H^- is a simple group or $H^- = \{1\}$.

6.13 Remarks

1. There are examples (cf. [BK] or Appendix [BT]) of uniform trees X for which $G = \mathrm{Aut}(X)$ is itself a free non-abelian discrete group, and hence $G^- = \{1\}$.

2. For any $H \leq G$, G_H (cf. (3.3)) satisfies $(I)_{X'}$, for all $X' \subset X$. Moreover G_H satisfies (LS) of (6.10), and so, by (6.10)(b),

$$(G_H)^- = (G_H)^Y := \left\langle \bigcup_{x \in VX, \deg(x) \geq 3} (G_H)_x \right\rangle,$$

and $G_H / (G_H)^Y$ is a free product of cyclic groups, each of order ∞ or 2.

3. Suppose that $X' \subset X_0 \subset X$ are subtrees, and X_0 is H-invariant. Then it follows directly from the definitions that: If H on X satisfies $(I)_{X'}$, then $H|X_0$ on X_0 satisfies $(I)_{X'}$.

4. For related simplicity results, see [BM3].

6.14   **Theorem.** *Let X be an infinite locally finite tree, and $\Gamma \leq H \leq G = \mathrm{Aut}(X)$ subgroups such that:*

(1) Γ *is discrete, and*

(2) $H = G_H$ *and acts minimally on X.*

   (a) *If H is discrete then $H \cong F * T$, where $F = \pi_1(H \backslash X)$ is a free group, and T is a free product of groups of order 2.*
   (b) *If H is discrete and Γ is an X-lattice then $[H : \Gamma] < \infty$, $H = C_H(\Gamma)$, and H, Γ and $K_H(\Gamma)$ are uniform X-lattices.*
   (c) *Assume that:*

(3) $\overline{C_H(\Gamma))} = H$ *and* $K_H(\Gamma) \neq \{1\}$.

If H fixes no end of X, or if Γ is an X-lattice, then H is discrete.

6.15   **Corollary** ("lattices with large commensurator"). *Let Γ be an X-lattice acting minimally on X. Suppose that $\Gamma \leq H = G_H \leq G$ and $\overline{C_H(\Gamma)} = H$. Then Γ is residually finite.*

Proof of (6.15). If $K_H(\Gamma) = \{1\}$ then Γ is residually finite by (6.1)(13). If $K_H(\Gamma) \neq \{1\}$ then (6.14)(c),(b), and (a) imply that Γ is virtually free, hence residually finite.

The following sections permit us to extend (6.15) to "parabolic lattices," which don't act minimally (cf. (6.18)).

Proof of (6.14). Part (a) follows from (3.5), as does part (b). We now prove (c). In view of (2), if H fixes no end of X then Theorem (6.12) applies, so $H^-(= H^Y$, by (6.10)) is a simple, and open, subgroup of H. Further, since (by assumption (3)) $\{1\} \neq K_H(\Gamma) \lhd \overline{C_H(\Gamma)} = H$, (6.12) tells us that $H^- \leq K_H(\Gamma) \leq \Gamma$, hence H^- is discrete, and hence H is discrete.

Suppose now that Γ is an X-lattice. If Γ fixes no end of X, then the previous discussion applies to show that H is discrete. Suppose, on the other hand, that Γ fixes an end ε of X. Then (see (5.13)) the length function ℓ for the Γ-action is abelian. If $\ell(\Gamma) = 0$ then, by Theorem (9.7) below, X is a "parabolic tree" (i.e., ε is its unique end), and the horoballs centered at E are all G-invariant, contradicting the minimally assumption in (2). Thus $\ell(\Gamma) \neq \{0\}$. In this case $X_\Gamma = X_G = X$, by (2), and so, by (5.15), X is a linear tree, so G, hence also H, is discrete.

6.16   Automorphism groups of rooted trees

A rooted tree (X, x_0) is defined by a sequence

(1) $$\cdots \to X_n \xrightarrow{t} X_{n-1} \to \cdots \to X_1 \xrightarrow{t} X_0 = \{x_0\},$$

where $X_n(\subset VX)$ is the sphere of radius n centered at x_0, and, for $x \neq x_0, t(x) = \partial_1(e_x)$, where e_x is the first edge of the reduced path from x to x_0.

(2)

Suppose that $H \leq G = \mathrm{Aut}(X)$ fixes x_0. Let $p : X \to A = H\backslash X$ be the quotient graph, and $a_0 = p(x_0)$. Then (A, a_0) is again a rooted tree, defined by

(3) $$\cdots \to A_n \xrightarrow{t} A_{n-1} \to \cdots \to A_1 \xrightarrow{t} A_0 = \{a_0\},$$

the quotient of (1) by the action of H. Let $(A, i) = I(H\backslash\backslash X)$.

For $x \in VX - \{x_0\}$, as in (2), we clearly have

(4) $$H_x = H_{e_x} \leq H_{t(x)}.$$

Putting

$$p\left(\begin{array}{ccc} \circ & \xrightarrow{} & \circ \\ x & e_x & t(x) \end{array} \right) = \left(\begin{array}{ccc} \circ & \xrightarrow{} & \circ \\ a & e_a & t(a) \end{array} \right)$$

we thus have

$$i(e_a) = 1, \quad \text{and we put}$$

(5)
$$q(a) := \Delta(e_a)(:= i(\bar{e}_a)/i(e_a)) = i(\bar{e}_a)$$

$$= [H_{t(x)} : H_x].$$

We also write, for $a \neq a_0$, and using the notation of (2.2)

(6)
$$A_0(a) = A_0(e_a) \quad \left(VA_0(a) = \bigcup_{n \geq 0} t^{-n}(a)\right), \quad \text{and}$$

$$A_0(a_0) = A.$$

Recall that $X = (\widetilde{A}, i, a_0)$, and $G_H = G_{(A,i)} = \{g \epsilon G \mid p \circ g = p\}$. For $a \epsilon VA$ we shall abbreviate

(7)
$$H(a) = G_{(A_0(a),i)}.$$

Thus

(8) $G_{(A,i)} = H(a_0)$, and we can naturally view $H(a)$ as a subgroup of $H(ta)$ for $a \neq a_0$.

With this notation, we can give the following inductive description of $G_{(A,i)}$.

(9)   For $a \epsilon VA$,

$$H(a) \cong \prod_{b \epsilon t^{-1}(a)} (H(b)^{q(b)} \rtimes S_{q(b)}).$$

For $b \epsilon t^{-1}(a)$, the inclusion $H(b) \leq H(a)$ indentifies $H(b)$ with one factor of $H(b)^{q(b)}$ ($\leq H(b)^{q(b)} \rtimes S_{q(b)}$).

Here $S_{q(b)}$ denotes the symmetric group on a set of cardinal $q(b)$, which, for our applications, we can take to be always finite, e.g., when X is locally finite. Clearly:

(10)   For $b \epsilon t^{-1}(a)$, the distinct conjugates of $H(b)$ in $H(a)$ have pairwise trivial intersection. If $q(b) > 1$ then $H(b)$ contains no non-trivial normal subgroup of $H(a)$.

By iteration of (9), and observing that $G_H = G_{(A,i)}$ is the inverse limit of it's restrictions to balls of increasing radius about x_0, we see that $G_{(A,i)} = H(a_0)$ is an infinite interated product of wreath products of symmetric groups (cf. [BORT], Ch. III).

6.17   Automorphism groups of ended trees

Let (X, ε) be an *ended tree*, with horosphere sequence.

(1)
$$\cdots \to X_{n-1} \xrightarrow{t} X_n \xrightarrow{t} X_{n+1} \to \cdots ,$$
$$\lim_{n \to \infty} X_n = \{\varepsilon\}.$$

Let $H \leq G = \mathrm{Aut}(X)$ be a subgroup fixing ε, with quotient $p : X \to A = H \backslash X$. Then (A, ε_A), $\varepsilon_A = p(\varepsilon)$, is again an ended tree, with horosphere sequence

(2)
$$\cdots \to A_{n-1} \xrightarrow{t} A_n \xrightarrow{t} A_{n+1} \to \cdots$$

the quotient of (1) by H. Put $(A, i) = I(H \backslash\backslash X)$. For $a \in VA$ we have the picture

(3)

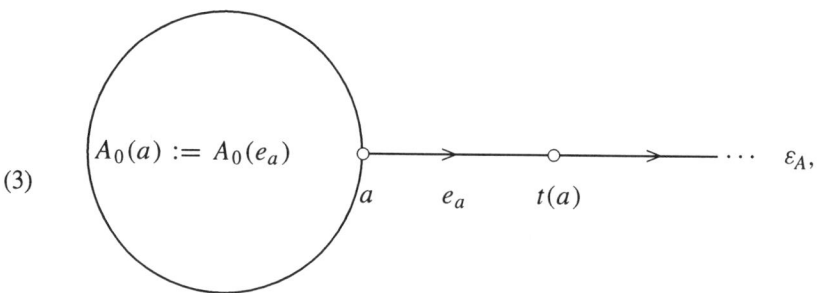

$$i(e_a) = 1, \quad \text{and} \quad q(a) = i(\bar{e}_a).$$

As in (6.16)(7), we put

(4)
$$H(a) := G_{(A_0(a), i)} \leq H(t(a)), \quad \text{so that}$$
$$G_{(A, i)} = \lim_{a \to \varepsilon_A} H(a).$$

Further, as in (6.16)(9), we have

(5)
$$H(a) \cong \prod_{b \in t^{-1}(a)} (H(b)^{q(b)} \rtimes S_{q(b)}).$$

From (4) and (6.16)(10) we conclude that:

(6)   Suppose that, for all $a \in VA$, $q(t^n a) > 1$ for some $n \geq 0$. Then $G_{(A, i)}$ contains no non-trivial discrete normal subgroups.

6.18   **Proposition** ("lattices with large commensurator"). *Let X be a locally finite tree and $\Gamma \leq H \leq G = \mathrm{Aut}(X)$ subgroups such that:*

(1)   *Γ is an X-lattice;*

(2) $H = G_H$; and

(3) $\overline{C_H(\Gamma)} = H$.

Then Γ is residually finite, except perhaps in the following situation:

(4)   *Γ is non-uniform, $\ell(\Gamma) \neq 0$, $\Gamma|X_\Gamma$ is residually finite, and $Ker(\Gamma \xrightarrow{res} \Gamma|X_\Gamma)$*
        is finite.

Remark. We know of no example as in (4) where Γ is not itself residually finite. In Chapters 9–10, we shall see examples of (necessarily non-uniform) tree lattices Γ which are not residually finite. It will follow then from (6.18) that $C_G(\Gamma)$ is not dense in G.

Proof of (6.18). Let $K = K_G(\Gamma)$, as in (6.1)(9). Then $K \leq \Gamma$, so K is discrete, and $K \lhd \overline{C_H(\Gamma))} = H$, by (6.2)(d) and assumption (3). If $K = \{1\}$ then Γ is residually finite, by (6.1)(13). So suppose that

(5) $K \neq \{1\}$.

If $\ell(\Gamma) = 0$ then (Theorem (9.7) below) X is a parabolic tree. Putting $(A, i) = I(H\backslash\backslash X)$, it follows from the fact that H contains a lattice that $\mathrm{Vol}(A, i) < \infty$. This implies the hypothesis of (6.17)(6), whence H ($= G_H$ (by (2)) $= G_{(A,i)}$) has no non-trivial discrete normal subgroups. Since K is a discrete normal subgroup of H, this contradicts (5). Thus we have:

(6) $\ell(\Gamma) \neq 0$.

Put $X' = X_\Gamma = X_G$ (by (5.11)) $= X_H$. For $L \leq G$ put $L' = L|X$; so that $G' \leq G^* = \mathrm{Aut}(X')$. From assumption (2), that $H = G_H$, it is easy to see that $H' \leq G^*$ satisfies the analogous condition: $H' = G^*_{H'} = \{g\epsilon G^*|p' \circ g = p'\}$, where $p' : X' \to H'\backslash X'$. Moreover since $\overline{C_H(\Gamma)} = H$ and $C_H(\Gamma)' \leq C_{H'}(\Gamma')$, we have $\overline{C_{H'}(\Gamma)} = H'$. Thus (Γ', H', X') satisfy the analogues of assumptions (1), (2), (3), as well as

(7) Γ' acts minimally on X'.

It follows then from (6.15) that Γ' is residually finite. Since Γ is discrete, $Ker(\Gamma \xrightarrow{res} \Gamma')$ is finite, thus completing the proof.

   To conclude, we recall some results and questions about tree lattice commensurators. We assume in what follows that X is a locally finite tree, and $\Gamma \leq G = \mathrm{Aut}(X))$ is a discrete subgroup.

6.19 **Uniform Density Theorem** (Yingsheng Liu [Li1]; cf. also [BK]). *If Γ is a uniform X-lattice then $C_G(\Gamma)$ is dense in G. Moreover $C(X) := C_G(\Gamma)$ depends, up to conjugacy, only on X, not Γ.*

6.20 **Question.** If Γ, Γ' are uniform X-lattices and $C_G(\Gamma) = C_G(\Gamma')$, must Γ and Γ' be commensurable?

6.21 Remarks

Suppose now that Γ is *not* a uniform X-lattice. What then can be said about the closure of $C_G(\Gamma)$? We list here various observations and examples.

(1)   If Γ is an X-lattice and $Ker(\Gamma \to \hat{\Gamma})$ is infinite, then $C_G(\Gamma)$ is not dense in G.

This follows from (6.18). Examples of Γ as in (1), in fact with $\hat{\Gamma} = \{1\}$, are given below in (9.11)(11) (with X a parabolic tree), and in (10.3) (with X a regular tree). Moreover (10.5) furnishes a non-uniform lattice on a tree X (of unbounded degree) such that $C_G(\Gamma') = N_G(\Gamma') = \Gamma$ with $|\Gamma/\Gamma'| = 2$, and G not discrete.

(2)   Let $F = F(S)$, a free group with basis S of cardinal d, and let X be the Cayley graph of $(F, S \cup S^{-1})$, so that $X = X_{2d}$, the $2d$-regular tree. Let $\Gamma = (F, F)$ (commutator subgroup). (Thus $\Gamma \backslash X$ is the Cayley graph of $(F^{ab}, S \cup S^{-1})$, the Euclidean lattice graph of $\mathbf{Z}^d \subset \mathbf{R}^d$.) Then $C_G(\Gamma)$ is dense in G.

This result is proved in [BM1], Prop. (8.1). Note that Γ acts freely on X, but it is not an X-lattice. The authors also furnish *the following example of a non-uniform tree lattice with dense commensurator* (cf. [BM1], Theorem (8.3)):

(3)   Let Γ, X be the fundamental group and universal cover of the following graph of groups,

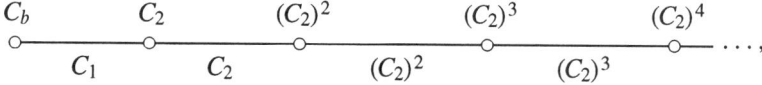

where C_m denotes a cyclic group of order m, and $b \geq 4$. Then $C_G(\Gamma)$ is dense in G.

The tree X in (3) is not uniform (cf. (10.1)(33)).

Recently, Shahar Mozes [M2] has shown that

(4)   The Nagao lattice $\Gamma = PSL_2(\mathbf{F}_q[t])$ on X_{q+1} (Example (10.2)) has dense commensurator.

Chapter 7

Existence of Tree Lattices

7.1   Introduction

Let X be a locally finite tree, and $G = \mathrm{Aut}(X)$. When does X admit a—uniform, or non-uniform—lattice? We would like to answer this in terms of properties of the edge-indexed quotient graph $I(G\backslash\backslash X)$ (cf. (2.4) and (2.5)).

More generally, consider a subgroup $H \leq G$ without inversions, quotient graph

$$p : X \to A = H\backslash X,$$

edge-indexed quotient

$$(A, i) = I(H\backslash\backslash X),$$

and group of deck transformations (cf. (3.3))

$$G_H = \{g \epsilon G \,|\, p \circ g = p\} = G_{(A,i)}.$$

We propose to seek conditions on (A, i) that characterize the existence of an X-lattice, uniform or non-uniform, contained in G_H. The following conditions are necessary (cf. (3.6)) for the existence of an X-lattice in G_H.

(U) (A, i) is unimodular.

(FV) $\mathrm{Vol}(A, i) < \infty$.

Conversely, we have the:

(1)   **Lattice Existence Theorem** (Appendix [BCR]).
        $[(\mathrm{U}) + (\mathrm{FV})] \Rightarrow [\exists \text{ an } X\text{-lattice } \Gamma \leq G_H]$.

This result was only conjectured, and proved under additional hypotheses, in an early draft of this book. A full proof was obtained by Lisa Carbone and Gabe Rosenberg, with help from the first author, during final proofs of the manuscript;

it appears here in Appendix [BCR]. We choose nonetheless to retain our original discussion since this provides techniques and examples of broader interest and which already have been used elsewhere (cf. [C1], [R]).

Recall from (2.6)(18) the condition

(BD) (A, i) has bounded denominators.

According to (3.7), (BD) is equivalent to the existence of a discrete $\Phi \leq G_H$ such that $\Phi \backslash X = H \backslash X$. Assuming (FV), Φ will be an X-lattice. Assuming

(F) A is finite,

Φ will be a uniform X-lattice. Thus we know from earlier results that

(2) $[(BD) + (FV)] \Rightarrow [\exists$ an X-lattice $\Gamma \leq G_H]$.

Further (cf. (3.10)),

$$
\text{(3)} \qquad\qquad \begin{aligned} [(BD) + (F)] &\Leftrightarrow [(U) + (F)] \\ &\Leftrightarrow [\exists \text{ a uniform } X\text{-lattice } \Gamma \leq H]. \end{aligned}
$$

The results (2) and (3) reduce the Lattice Existence Theorem to the case when (A, i) is (infinite) without bounded denominators. Using a process called "open fanning," we shall establish the result in this chapter in a large number of cases (cf. (7.4), (7.5), (7.6)). Theorem (7.4), in particular, covers the case when A is homologically of finite type, for example a tree.

In the setting of (3), when A is finite, and G_H (in fact, H itself) contains uniform X-lattices, we can further ask whether G_H also contains non-uniform X-lattices. The answer is negative in general, for at least the following reason.

(4)   If A is finite and $G_H = G_{(A,i)}$ is discrete, then G_H itself is a uniform X-lattice, so it cannot contain a non-uniform X-lattice. In fact, whenever G_H is an X-lattice, it must be uniform, i.e., A is finite. (See Appendix [BT], (1.1).)

When $G_H = G_{(A,i)}$ is discrete (see Appendix [BT] for a criterion for this to happen), then we say that (A, i) is *rigid*. More generally, if there is a G_H-invariant subtree X_0 such that $G_H | X_0$ is discrete, then we say that (A, i) is *virtually rigid*. Put $H_0 = H | X_0 \leq G_0 = \text{Aut}(X_0)$, and $A_0 = H \backslash X_0 = H_0 \backslash X_0 \subset A$, so that $(A_0, i) = I (H_0 \backslash\backslash X_0)$. It is not difficult to show that

$$G_H | X_0 = (G_0)_{H_0} = (G_0)_{(A_0, i)}.$$

Hence, if (A, i) is virtually rigid relative to X_0, as above, then (A_0, i) is rigid.

Suppose, as above, that (A, i) is virtually rigid, and that G_H contains an X-lattice Γ. Then $\Gamma_0 = \Gamma | X_0$ is an X_0-lattice in the discrete group $G_H | X_0 = (G_0)_{H_0}$, so the latter is also an X_0-lattice of finite index over Γ_0. It follows from Appendix [BT], (1.3) (see also (3.5) above) that $G_H | X_0$, hence also Γ_0, is a uniform X_0-lattice. Now it follows from (5.10) that Γ is also a uniform X-lattice. Conclusion: If (A, i) is virtually rigid, then $G_{(A,i)}$ cannot contain a non-uniform X-lattice.

We originally speculated that virtual rigidity of (A, i) represents the only obstruction to the existence of non-uniform lattices on uniform trees. This has since been proved by Lisa Carbone [C1].

(5)   **Theorem** (Non-uniform lattices on uniform trees [C1]). *Assume that:*

   (U) (A, i) *is unimodular.*

   (F) A *is finite.*

   (NVR) (A, i) *is not virtually rigid.*

   Then $G_{(A,i)}$ contains a non-uniform X-lattice.

We had earlier verified this result in several special cases. These are presented in Chapter 8 since Carbone makes use of them in her proof.

In our discussions below, the focus will be on (A, i). In fact, given only (A, i), then, choosing a base point $a_0 \epsilon VA$, we can construct $X = \widehat{(A, i, a_0)}$ and $G = \text{Aut}(X)$, and the projection $p : X \to A$. The latter defines (A, i) since, if $x \epsilon VX$, $a = p(x)$, $p_{(x)} : E_0(x) \to E_0(a)$, and $e \epsilon E_0(a)$, then $i(e) = |p_{(x)}^{-1}(e)|$. We can choose some faithful (e.g., cyclic) grouping $\mathbf{A}$ of (A, i), take $H = \pi_1(\mathbf{A}, a_0) \leq G$, and we then have G_H. In fact, we can directly write $G_H = G_{(A,i)} = \{g \epsilon G| \ p \circ g = p\}$, without reference to any particular choice of H.

It is intrinsic to this method, using only (A, i), that the X-lattices Γ we produce are only known to be contained in G_H, not in H itself. In the case of uniform X-lattices Γ, using the fact that Γ is virtually free, together with the Conjugacy Theorem (3.4), one is able to obtain lattices actually contained in H (cf. (3.10)). In the case of non-uniform lattices however there is no general method that accomplishes the same purpose; see the example in (6) below. (See [L3] for a treatment of lattices in simple rank 1 Lie groups over local fields of characteristic $p > 0$.)

(6)   **Example.** Let $X = X_{p+1}$, p prime, and $H = PGL_2(\mathbf{Q}_p) \leq G = \text{Aut}(X)$. Then $I(H \backslash\backslash X) = {}_\circ\overline{\quad p+1 \qquad\qquad p+1\quad}_\circ$, and G_H has index 2 in G. Moreover, G_H contains non-uniform X-lattices, e.g., $\Gamma = PGL_2(\mathbf{F}_p[t])$ (cf. (10.2) below), whereas all lattices in H are uniform (cf. (0.5)(1) above).

7.2   Open fanning

Let (A, i) be a (connected, as always) edge-indexed graph. Let $e \epsilon EA$, $\partial_k e = a_k$ ($k = 0, 1$). As in (2.1), we call e a *separating edge* if $A - \{e, \bar{e}\}$ is no longer connected. Let $A_k = A_k(e)$ denote the connected component containing a_k. Putting $m_0 = i(e)$ and $m_1 = i(\bar{e})$, we then have the schematic picture.

(1) $(A, i) = $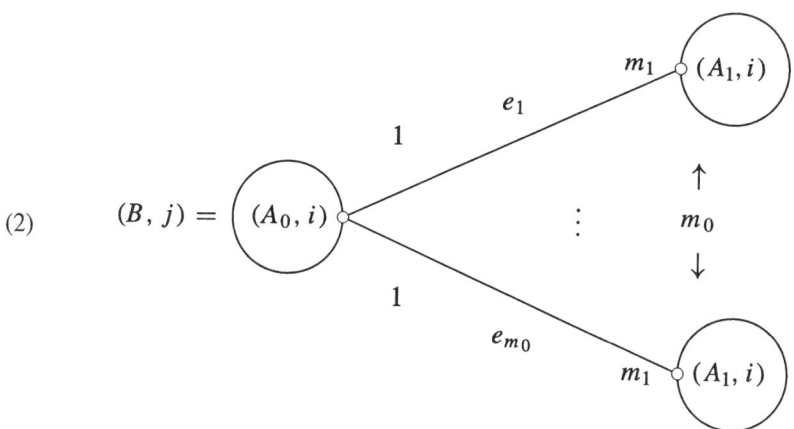

The "*open fanning of e in* (A, i)" is the edge-indexed covering graph (B, j) obtained
by replacing e, with index $i(e) = m_0$, by m_0 replicas of e, each (unramified) with
index 1, and changing nothing else. Thus,

(2) $(B, j) = $

There is an evident covering $p : (B, j) \to (A, i)$ (cf. (2.5)(5)). Moreover it is
clear that (B, j) is unimodular iff (A, i) is so, and it is easily seen that

(3) $\mathrm{Vol}_{a_0}(B, j) = \mathrm{Vol}_{a_0}(A, i)$

Suppose next that A is a tree, i.e., that every edge is separating. Fix a base point
$a_0 \epsilon VA$. Let E^+A consist of all edges $e \epsilon EA$ which are directed away from a_0. Then
we can simultaneously open fan all of the $e \epsilon E^+A$. Visualize this as starting with the
e's adjacent to a_0, $e \epsilon E_0(a_0)$, and then proceeding inductively to edges at increasing
distance from a_0. The result is a covering of rooted edge-indexed trees,

(4) $p : (B, j, b_0) \to (A, i, a_0)$,   with finite fibers.

Moreover

(5) $\mathrm{Vol}_{b_0}(B, j) = \mathrm{Vol}_{a_0}(A, i)$

and (B, j, b_0) is "non-decreasing," in the sense that $j(e) = 1$ for all $e \epsilon EB$ directed
away from b_0. If $\gamma = (e_1, \ldots, e_n)$ is a reduced edge path from $b_0 = \partial_0 e_1$ to $b = \partial_1 e_n$ in B, and if $e \epsilon E_0(b)$, then $\triangle(\gamma)/i(e) = i(\bar{e}_1) \ldots i(\bar{e}_n)/i(e))$, which equals

$i(\bar{e}_1)) \ldots i(\bar{e}_n)$ if $e \neq \bar{e}_n$, and $i(\bar{e}_1) \ldots i(\bar{e}_{n-1})$ if $e = \bar{e}_n$. In either case, $\Delta(\gamma)/i(e) \epsilon \mathbf{Z}$. Hence (cf. (2.6)(18)),

(6) (B, j) has bounded denominators.

Let $c = p(b)$, the endpoint of $p(\gamma) = (p(e_1), \ldots, p(e_n)) = (f_1, \ldots, f_n)$. We have $i(\bar{f}_h) = j(\bar{e}_h)$, while $j(e_h)) = 1$ $(h = 1, \ldots, n)$. Thus $\frac{\Delta b}{\Delta b_0} = j(\bar{e}_1) \ldots j(\bar{e}_n) = i(\bar{f}_1) \ldots i(\bar{f}_n) = \frac{\Delta c}{\Delta a_0} \cdot i(f_1) \ldots i(f_n) = \frac{\Delta c}{\Delta a_0} |p^{-1}(c)|$, clearly. Thus,

(7)   For $b \epsilon VB$ and $c = p(b) \epsilon VA$, $\frac{\Delta b}{\Delta b_0} = \frac{\Delta c}{\Delta a_0} |p^{-1}(c)|$, so

$$\mathrm{Vol}_b(B, j) = \mathrm{Vol}_c(A, i) \cdot |p^{-1}(c)|.$$

7.3   Example. Let

(1) $(A, i, a_0) =$

Then $\frac{\Delta a_n}{\Delta a_0} = (3/2)^n$, so

(2) $\mathrm{Vol}_{a_0}(A, i) = \sum_{n \geq 0} (2/3)^n = 3 < \infty,$

whereas

(3) (A, i) does not have bounded denominators.

Since A is a tree, (A, i) is unimodular. Thus, in the notation of (7.1),

(4) (A, i) satisfies (U) and (FV), but not (BD).

Hence G contains no discrete subgroup Φ such that $\Phi \backslash X = A$, even though there are non-discrete subgroups with this property.

According to the Lattice Existence Theorem in (7.1), $G_{(A, i)}$ contains an X-lattice. To see here that this is so, consider the "open fanning" (B, j, b_0) of (A, i, a_0), as in (7.2), with the covering $p : (B, j, b_0) \rightarrow (A, i, a_0)$.

(5) $(B, j) = b_0$

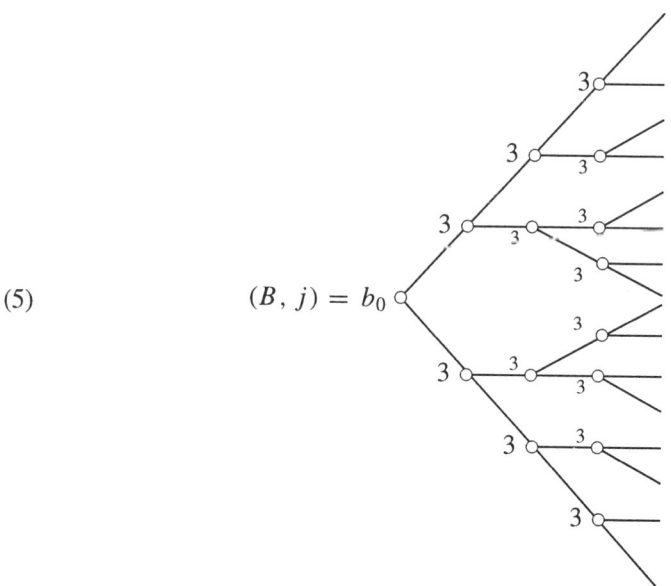

We have

$$\mathrm{Vol}_{b_0}(B, j) = \mathrm{Vol}_{a_0}(A, i) = 3 \qquad < \infty,$$

and (B, j) has bounded denominators. Hence, putting $X = (\widetilde{A, i, a_0}) = (\widetilde{B, j, b_0})$, there is a discrete $\Phi \leq G$ such that $I(\Phi \backslash\backslash X) = (B, j)$, and so Φ is a (non-uniform) X-lattice. Moreover $\Phi \leq G_{(B,j)} \leq G_{(A,i)}$, thus confirming the Lattice Existence Theorem (7.1)(1) in this case.

7.4   Theorem. *Let X be a locally finite tree, $H \leq G = \mathrm{Aut}(X)$ a subgroup without inversions, and $(A, i) = I(H \backslash\backslash X)$. Assume:*

(U)   *(A, i) is unimodular;*

(FV)   *$\mathrm{Vol}(A, i) < \infty$; and*

(HFT)   *A is homologically of finite type, or equivalently, $\pi_1(A)$ is finitely generated.*

Then there is an X-lattice $\Gamma \leq G_H$ with the following properties, where

$$p : B := \Gamma \backslash X \longrightarrow A = H \backslash X = G_H \backslash X.$$

is the natural projection.

(a) *p has finite fibers, so Γ is a uniform G_H-lattice.*

(b) *Let $b \in VB$ and $a = p(b)$. Then $\pi_1(B, b) \to \pi_1(A, a)$ is an isomorphism, and*

$$\mathrm{Vol}_b(I(\Gamma \backslash\backslash X)) = \mathrm{Vol}_a(I(H \backslash\backslash X)) \cdot |p^{-1}(a)|.$$

Proof. Since A is homologically of finite type, there is a finite connected subgraph A_0 of A that supports all of the homology. Then A is obtained from A_0 by attaching, at each $a \in VA_0$, a rooted tree (A_a, a)

(1)

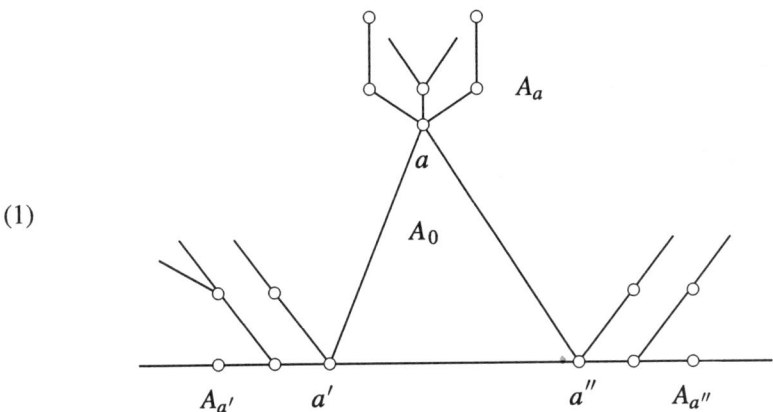

Clearly unimodularity of (A, i) is equivalent to that of (A_0, i). Since (A_0, i) is unimodular, bounded denominators for (A, i) is equivalent to it's validity for each (A_a, i). Theorem (3.7) covers the case when (A, i) has bounded denominators, so assume that this is not the case.

For each $a \in VA$, let

(2) $$p_a : (B_a, j_a, b_a) \to (A_a, i, a)$$

denote the covering by the open fanning of (A_a, i, a), as in (7.2)(4). Thus, (B_a, b_a) is a rooted tree, p_a has finite fibers,

(3) $$\mathrm{Vol}_{b_a}(B_a, j_a) = \mathrm{Vol}_a(A_a, i),$$

and (B_a, j_a) has bounded denominators (cf. (7.2)(4), (5), and (6)).

Now let

(4) $$(B, j) = (A_0, i) \cup \bigcup_{a \in VA_0} (B_a, b_a),$$

obtained from (A, i) by removing (A_a, i) at $a \in VA$, and replacing it with (B_a, j_a), attached by indentifying $b_a \in VB_a$ with $a \in VA_0$. Define

(5) $$p : (B, j) \to (A, i)$$

to be the identity on A_0, and p_a on B_a. Then clearly p is a covering with finite fibers inducing an isomorphism on fundamental groups, (B, j) is unimodular, and, in fact, (B, j) has bounded denominators, since each (B_a, j_a) does, and there are only finitely many of these (A_0 is finite). Finally, to calculate volumes, let $a_0 \epsilon VA_0$. Then for $b \epsilon VB_a$, $a \epsilon VA_0$, we have, in (B, j)

(6)
$$\frac{\Delta b}{\Delta a_0} = \frac{\Delta a}{\Delta a_0} \cdot \frac{\Delta b}{\Delta a},$$

where the first factor is calculated in (A_0, i), and the second in (B_a, j_a). Thus

(7)
$$\mathrm{Vol}_{a_0}(B, j) = \sum_{a \epsilon VA_0} \frac{\Delta a_0}{\Delta a} \cdot \mathrm{Vol}_{b_a}(B_a, j_a)$$

$$\overset{(3)}{=} \sum_{a \epsilon VA_0} \frac{\Delta a_0}{\Delta a} \mathrm{Vol}_a(A_a, i)$$

$$= \mathrm{Vol}_{a_0}(A, i).$$

From this and (7.2)(7) it follows that, if $b \epsilon VB$ and $c = p(b)$, then

(8)
$$\mathrm{Vol}_b(B, j) = \mathrm{Vol}_c(A, i) \cdot |p^{-1}(c)|.$$

Since (B, j) has bounded denominators it follows from Theorem (3.7) that there is a discrete $\Gamma \leq G_{(B,j)} \leq G_{(A,i)} = G_H$, such that $I(\Gamma \backslash\backslash X) = (B, j)$. Since $\mathrm{Vol}(B, j) < \infty$ (by (7) and (FV)), Γ is an X-lattice. For $x \epsilon VX$ and $a = p_H(x) \epsilon VA$, we have

$$p^{-1}(a) = \Gamma \backslash (G_H \cdot x) \cong \Gamma \backslash G_H / (G_H)_x.$$

Since p has finite fibers and $(G_H)_x$ is compact, it follows that Γ is a uniform G_H-lattice (cf. (1.5)(8)). This establishes (a), and (b) from (8) above.

7.5 Multiple open fanning

We indicate here a generalization of open fanning that suffices to prove an analogue of Theorem (7.4) even in some cases when A has infinitely generated homology.

Let (A, i) be an edge indexed graph. A subset $E \subset EA$ will be called an *(oriented) bridge* of A if $E \cap \bar{E} = \phi$, and $A - (E \cup \bar{E}) = A_0 \amalg A_1$, with A_0 and A_1 connected subgraphs, and $\partial_k e \epsilon A_k$ $(k = 0, 1)$ $\forall e \epsilon E$. We call E a *bridge from A_0 to A_1*.

(1)

We further call E a *bridge of (A, i), of index m*, if $i(e) = m$ $\forall e \epsilon E$.

Under these conditions, the "open fanning of E in (A, i)" is the covering $p :$ $(B, j) \to (A, i)$ defined by

(2) $(B, j) = (A_0, i)$

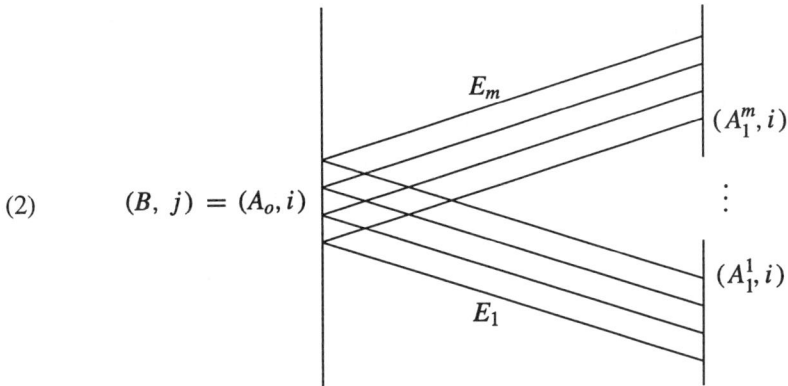

Precisely, $(B, j) = (A_0, i) \cup \bigcup_{h=1}^{m} [E_h \cup \bar{E}_h \cup (A_1^h, i)]$, where each $E_h \cup \bar{E}_h \cup (A_1^h, i)$ is a copy of $E \cup \bar{E} \cup (A_1, i)$, attached to A_0 as given, and, for $e_h \epsilon E_h$ corresponding to $e \epsilon E$, we have $j(\bar{e}_h) = i(\bar{e})$ and $j(e_h) = 1$. The covering p is the obvious one.

When $E = \{e\}$, a single edge, this construction reduces to that of (7.2). As in that case,

(3)   If (A, i) is unimodular then so also is (B, j).

In the present case this requires a small verification, which is not difficult, using the fact that $i(e) = m \;\; \forall \, e \epsilon E$. Further, just as in (7.2), one sees easily that:

(4)   For $a_0 \epsilon VA_0$, $\mathrm{Vol}_{a_0}(B, j) = \mathrm{Vol}_{a_0}(A, i)$.

Now suppose that we have a "*tree of connected graphs.*" By this we mean a triple (A, T, q), where A is a connected graph, T is a tree, and

$$q : VA \cup EA \to VT \cup ET$$

is a (surjective) map such that,

(i) $\forall \, t \epsilon VT$, $A_t := q^{-1}(t)$, is a connected subgraph of A; and

(ii) $\forall \, e \epsilon ET$, $E_e := q^{-1}(e)$ is an oriented bridge from $A_{\partial_0 e}$ to $A_{\partial_1 e}$, and $E_{\bar{e}} = \bar{E}_e$.

Suppose now further that (A, i) is an edge-indexed graph. Let $\bullet \epsilon VT$ be a base point, and let $E^+ T$ denote the set of $e \epsilon ET$ which are directed away from $\bullet$. We further assume,

(iii) $\forall \, e \epsilon E^+ T$, E_e is a bridge of (A, i), i.e., $i(f)$ is constant for $f \epsilon E_e$.

Under these conditions we can open fan each of the E_e ($e \epsilon E^+ T$), say starting with $e \epsilon E_0(\bullet)$, and then proceeding by increasing distance from $\bullet$ as in the latter part of (7.2). The final result is a covering,

(5) $p : (B, j) \to (A, i).$

Further, just as above,

(6)   If (A, i) is unimodular then so also is (B, j),

and

(7)   For $a_0 \in VA_\bullet$, $\mathrm{Vol}_{a_0}(B, j) = \mathrm{Vol}_{a_0}(A, i)$.

Finally, and this is the main point,

(8)   If the (A_t, i) $(t \in VT)$ have "uniformly bounded denominators," then (B, j) has bounded denominators.

Thus we can choose a faithful finite grouping of (B, j) to produce a discrete fundamental group $\Gamma \leq G_{(B,j)} \leq G_{(A,i)}$, which, if $\mathrm{Vol}(A, i) < \infty$, furnishes the desired tree lattice.

Suffice it to illustrate this construction with some simple examples.

7.6   **Example.** 1. Consider

A projects to the tree

with

Then the open fanning $p : (B, j) \to (A, i)$ has the following form

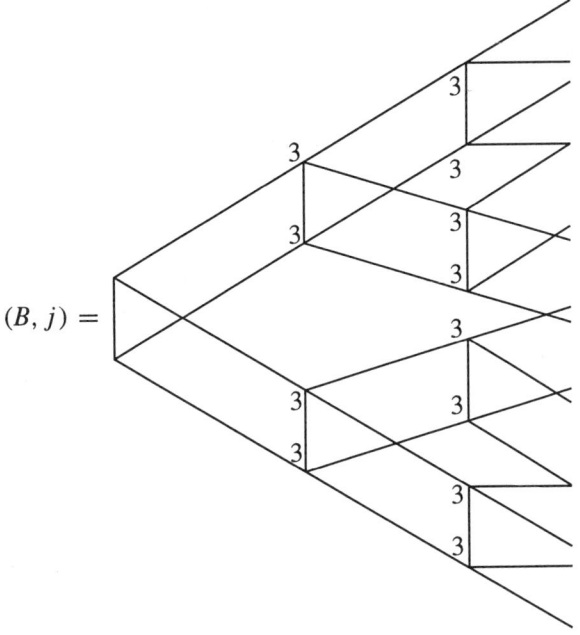

$$(B, j) =$$

In this example, (A, i) is unimodular and has finite volume, but unbounded denominators, whereas (B, j) is unimodular, has finite volume, and bounded denominators. This example is not covered by (7.4) since A has infinitely generated homology.

?   Consider $A = \mathbf{Z}^2$, the Euclidean plane lattice grid, and edges indexed $\overset{2}{\circ} \longrightarrow \overset{3}{\circ}$ when they move away from the origin (horizontally or vertically). Thus, (A, i) is clearly unimodular with infinitely generated homology, and, if $\sigma = \sum_{n \geq 1}(2/3)^n = 2$,

$$\text{Vol}_{(0,0)}(A, i) = \sum_{m,n \geq 0} \left(\frac{2}{3}\right)^{|m|+|n|} = \left(\sum_{n \geq 0} \left(\frac{2}{3}\right)^{|n|}\right)^2$$
$$= (1 + 2\sigma)^2 = 25 < \infty.$$

Now projection to the y-axis expresses (A, i) as an infinite line of graphs

$$(A_n, i) = \cdots \overset{2}{\underset{}{\longrightarrow}} \overset{3}{\underset{\circ}{}} \overset{2}{\underset{}{}} \overset{3}{\underset{\circ}{}} \overset{2}{\underset{}{}} \overset{2}{\underset{\bullet}{}} \overset{3}{\underset{}{}} \overset{2}{\underset{\circ}{}} \overset{3}{\underset{}{}} \overset{2}{\underset{\circ}{}} \cdots$$
$$\text{origin}$$

connected by (infinite) bridges of the form

$$(E_n, i) = \left\{ \cdots \quad \begin{matrix} \circ & \circ & \circ & \circ & \circ \\ 3\big\uparrow & 3\big\uparrow & 3\big\uparrow & 3\big\uparrow & 3\big\uparrow \\ 2\big\uparrow & 2\big\uparrow & 2\big\uparrow & 2\big\uparrow & 2\big\uparrow \\ \circ & \circ & \circ & \circ & \circ \end{matrix} \quad \cdots \right\}$$

if $n \geq 0$, or, if $n < 0$, the edges are directed downward.

The covering

$$p : (B, j) \to (A, i)$$

of (7.5) is obtained from (A, i) by bifurcating edges that move (vertically) away from the x-axis. We can visualize (B, j) as embedded in 3-space, the bifurcations occurring in planes parallel to the (y, z)-plane. What (B, j) has achieved is to unramify the edges parallel to the y-axis that move away from the x-axis. However, since the (A_n, i)'s have unbounded denominators, so does (B, j). To repair this, consider planes in 3-space defined by: (x-coordinate) $= n + \frac{1}{2}$ ($n \epsilon \mathbf{Z}$). Each such plane bisects a set E_n of oriented edges of (B, j) that move away from the (y, z)-plane. These sets E_n are bridges of (B, j), joining connected subgraphs (B_n, j) of (B, j) which look like two copies of example (7.3)(5), joined at the root. In particular these have uniformly bounded denominators. This writes (B, j) as an infinite line of these connected (B_n, j)'s. If we then open fan (B, j) according to this decomposition, we obtain a covering $(C, k) \to (B, j)$ with (C, k) unimodular, of finite volume, and with bounded denominators. Thus, again, we succeed in producing a tree lattice $\Gamma \leq G_{(A, i)}$.

Chapter 8

Non-Uniform Lattices on Uniform Trees

8.1 Carbone's Theorem

Consider a locally finite tree X, a subgroup

$$H \leq G = \text{Aut}(X)$$

without inversions, and indexed quotient graph

$$(A, i) = I(H \backslash\backslash X).$$

We here discuss and prove special cases of Carbone's Theorem $(7.1)(5)$.

Theorem ([C1, 3]). *Assume that:*

(U)   *(A, i) is unimodular.*

(F)   *A is finite.*

(NVR)   *(A, i) is not virtually rigid.*

Then $G_H = G_{(A,i)}$ contains a non-uniform X-lattice.

We know from Theorem (3.10) that, from conditions (U) and (F) above, there is a free *uniform* X-lattice $\Gamma \leq H$, and a *uniform* X-lattice $\Phi \leq G_H$ such that $\Phi \backslash X = A$. Carbone's theorem says that, with the additional condition (NVR), there exists a *non-uniform* lattice $\Gamma \leq G_H$ as well. It is shown after $(7.1)(4)$ that, if (A, i) is virtually rigid, then $G_{(A,i)}$ cannot contain a non-uniform X-lattice.

We shall here treat special cases of Carbone's Theorem, under the additional assumption

(Min)   *H acts minimally on X.*

In fact, Carbone first treats this (main) case [C1] and then deduces the general case from this one [C3]. In the presence of (Min), condition (NVR) is equivalent to

(NR) (A, i) is not rigid.

In turn, it follows from Theorem (3.5), (3.5)(1)$_{min}$ in particular, that, in our context, (NR) can be equivalently replaced by the condition:

(NDR)   For some $e \epsilon EA$, either $i(e) \geq 3$ or $i(e) = 2$ and $E_0(\partial_0 e) \neq \{e\}$.

Since, under conditions (U) + (F), there exist X-lattices $\Gamma \leq H$, it follows from (5.12) that H acts minimally iff every X-lattice acts minimally, iff G acts minimally, and iff X has no terminal vertices. (A vertex $x \epsilon VX$ is *terminal* if $\deg(x)$ $(= |E_0(x)|) = 1)$. The image in A of a terminal vertex of X will be a *terminal vertex of* (A, i), and conversely. For $a \epsilon VA$ we put

(1)
$$\deg_{(A,i)}(a) = \sum_{e \epsilon E_0(a)} i(e)$$

and call a *a terminal vertex of* (A, i) if $\deg_{(A,i)}(a) = 1$. In conclusion, we can restate condition (Min) above in the form:

(Min) (A, i) has no terminal vertices, i.e., $\deg_{(A,i)}(a) > 1 \; \forall \; a \epsilon VA$.

Finally, what does it mean to have a non-uniform lattice $\Gamma \leq G_H$? Put

$$(B, j) = I(\Gamma \backslash\backslash X).$$

Then we have a *covering* (cf. (2.5)(4))

$$p : (B, j) = I(\Gamma \backslash\backslash X) \to (A, i) = I(G_H \backslash\backslash X),$$

that reflects the inclusion $\Gamma \leq G_H$. Since (B, j) admits the finite grouping $\Gamma \backslash\backslash X$ it follows from (2.6)(19) that

(BD) (B, j) has bounded denominators.

Further, since Γ is an X-lattice,

(FV) $\text{Vol}(B, j) < \infty$.

Finally, the fact that Γ is non-uniform is expressed by the fact that

(Inf) B is infinite.

Now suppose conversely that we can produce an infinite, finite volume, edge-indexed graph (B, j) with bounded denominators, with a covering $p : (B, j) \to (A, i)$. Then by (2.6)(19), we can find a faithful finite grouping $\mathbf{B}$ of (B, j). Taking

$\Gamma = \pi_1(\mathbf{B}, b)$ $(b \epsilon VB)$, we will have found the desired non-uniform X-lattice $\Gamma \leq$ $G_{(B,j)} \leq G_{(A,i)} = G_H$. We now summarize this:

(2)   **Reformulation of Carbone's Theorem; Minimal Case.** *Assume that:*

  (U) (A, i) *is unimodular.*

  (F) A *is finite.*

(NDR) $\exists\, e \epsilon EA$ *such that either* $i(e) \geq 3$, *or* $i(e) = 2$ *and* $E_0(\partial_0 e) \neq \{e\}$.

  (Min) (A, i) *has no terminal vertices, i.e.,* $i(e) > 1$ *if* $\partial_0 e$ *is a terminal vertex of* A.

Then there exists a covering $p : (B, j) \rightarrow (A, i)$, *satisfying that*

  (BD) (B, j) *has bounded denominators;*

  (FV) $\mathrm{Vol}(B, j) < \infty$; *and*

  (Inf) B *is infinite.*

In this section, we shall prove this result in a variety of special cases that were the starting point for Carbone's work and that illustrate some of the methods used. The following theorem covers some (but by no means all) of these cases and is chosen for ease of description. We give its group theoretic formulation, noting that the four lead hypotheses of (8.2) are equivalent to those of (8.1)(2).

In what follows we shall adopt the following *notational convention:*

(3)   For $n \epsilon \mathbf{Z}$ we write $n' = n - 1$.

8.2   **Theorem.** *Let X be a locally finite tree, $H \leq G = \mathrm{Aut}(X)$ a subgroup without inversions, and $(A, i) = I(H \backslash\backslash X)$. Assume:*

  (U) (A, i) *is unimodular;*

  (F) A *is finite;*

  (ND) G_H *is non-discrete; and*

  (Min) H *acts minimally on X.*

Then G_H contains a non-uniform X-lattice provided that either A is a tree, or A contains a separating edge e with $i(e)' \cdot i(\bar{e})' > 1$ (cf. (8.1)(3)).

The proof will be given in (8.6) below, based on cases exhibited in the following examples. We also establish Theorem (8.1) when $|VA| \leq 2$, using a construction furnished by Lisa Carbone.

8.3   **Example.** In this and the following examples, (A, i) will always be understood to be an edge-indexed graph satisfying the conditions (U), (F), (NDR), and (Min) of (8.1)(2).

(a)
$$(A, i) = \left((A_0, i) \underset{a_0}{\overset{m_0}{\bigcirc}} \quad\quad \underset{a_1}{\overset{m_1}{\bigcirc}} (A_1, i) \right),$$

(b) $(A, i) = $

We seek a covering $p : (B, j) \to (A, i)$ satisfying conditions (BD) (bounded denominators), (FV) ($Vol(B, j) < \infty$), and (Inf) (B is infinite) of (8.1)(2).

In case (a) (assuming $m_0' m_1' > 0$) we take $(B, j) =$

(1)

The edge-indices on each copy of A_0 and A_1 are as given in (A, i). The covering $p : (B, j) \to (A, i)$ is fairly obvious ($a_n \epsilon VB$ maps to $a_{(n \bmod 2)}$ in A). It is clear that (B, j) is unimodular and infinite. Moreover one see immediately that (cf. (2.6)(7)),

(2) $$\frac{\Delta a_{2n}}{\Delta a_0} = \frac{(m_0' m_1')^n}{m_0}, \qquad \frac{\Delta a_{2n+1}}{\Delta a_0} = \frac{\Delta a_{2n}}{\Delta a_0} \cdot m_1'.$$

From (2) it follows easily that (B, j) has bounded denominators, and

(3) $$\mathrm{Vol}(B, j) < \infty \quad \text{if } m_0' m_1' > 1.$$

In case (b) (assuming $m_0' m_1' > 0$) we take

$(B, j) =$

(4)

Again, (B, j) is clearly infinite and unimodular, and the covering $p : (B, j) \to (A, i)$ is obvious. Let $a_0, a_1, a_2, a_3, a_4, \ldots$ denote the successive vertices at which the copies of A_0 and A_1 are attached, and $d_1, d_2, d_3, d_4, \ldots$, the successive vertices of the copies of C. Note then that $p(d_1), p(d_2), p(d_3), p(d_4), p(d_5), p(d_6), \ldots = c_0, c_1, c_1, c_0, c_0, c_1, c_1, c_0, \ldots$. It follows that

$$\frac{\Delta a_{2n}}{\Delta a_0} = \left(\frac{r_0}{m_0} \frac{\Delta c_1}{\Delta c_0} \frac{m_1'}{r_1} \frac{r_1}{1} \frac{\Delta c_0}{\Delta c_1} \frac{m_0'}{r_0} \right) \left(\frac{r_0}{1} \frac{\Delta c_1}{\Delta c_0} \frac{m_1'}{r_1} \frac{r_1}{1} \frac{\Delta c_0}{\Delta c_1} \frac{m_0'}{r_0} \right)$$
$$\cdots \left(\frac{r_0}{1} \frac{\Delta c_1}{\Delta c_0} \frac{m_1'}{r_1} \frac{r_1}{1} \frac{\Delta c_0}{\Delta c_1} \frac{m_0'}{r_0} \right).$$

Since $\frac{\Delta c_1}{\Delta c_0} \frac{\Delta c_0}{\Delta c_1} = 1$ we have

(5) $$\frac{\Delta a_{2n}}{\Delta a_0} = \frac{(m_0' m_1')^n}{m_0}$$

and

(5')
$$\frac{\Delta a_{2n+1}}{\Delta a_0} = \frac{\Delta a_{2n}}{\Delta a_0} \cdot \frac{r_0}{1} \frac{\Delta c_1}{\Delta c_0} \frac{m_1'}{r_1}.$$

It follows easily from (5) and (5') that (B, j) has bounded denominators and

(6) $\mathrm{Vol}(B, j) < \infty$ if $m_0' m_1' > 1.$

Conclusion

(7)   If, in (a) or (b) above, we have $m_0' m_1' > 1$, then (A, i) admits a covering
        $p : (B, j) \to (A, i)$ satisfying (BD), (FV), and (Inf).

8.4   Example. Suppose, in Example (8.3), that

(1) $\pi_1(A_1) \neq \{1\}.$

Since this fundamental group is free, it has an index 2 subgroup, whence A_1 has a
double cover

(2) $q : C_1 \to A_1.$

Since q is locally bijective, we can form an edge-indexed graph (C_1, i) so that q
is index preserving. Since (A_1, i) is unimodular, so also is (C_1, i). Put $q^{-1}(a_1) = \{c_0, c_1\}$. It then further follows from unimodularity of (A_1, i) that

(3)
$$\frac{\Delta c_0}{\Delta c_1} = 1 \text{ in } (C_1, i) = c_0 \mathbin{\text{\huge ◯}} c_1.$$

Now consider

(a) $(A', i) = \widehat{A_0} \overset{m_0}{\underset{a_0}{\rule{1.2cm}{0.4pt}}} \overset{m_1}{\underset{c_0}{\rule{1.2cm}{0.4pt}}} C_1 \overset{m_1}{\underset{c_1}{\rule{1.2cm}{0.4pt}}} \overset{m_0}{\underset{a_0}{\rule{1.2cm}{0.4pt}}} \widehat{A_0}$, respectively,

(b) $(A', i) = \widehat{A_0} \overset{m_0}{\underset{a_0}{\rule{1cm}{0.4pt}}} \overset{r_0}{\rule{1cm}{0.4pt}} C \overset{r_1}{\underset{c_0}{\rule{1cm}{0.4pt}}} \overset{m_1}{\rule{1cm}{0.4pt}} C_1 \overset{m_1}{\underset{c_1}{\rule{1cm}{0.4pt}}} \overset{r_1}{\rule{1cm}{0.4pt}} C \overset{r_0}{\rule{1cm}{0.4pt}} \overset{m_0}{\underset{a_0}{\rule{1cm}{0.4pt}}} \widehat{A_0}$.

Then, using the covering q in (2) above, we obtain a covering $q : (A', i) \to (A, i)$ of
the Examples (8.3) (a) and (b), respectively. If we now further assume that

(4) $m_0 \geq 3,$

then, since $m_0' m_0' \geq 4 > 1$, the conditions of (8.3)(7) apply to (A', i).

Conclusion

(5)　　If, in (8.3)(a) or (b), we have $m_0 \geq 3$ and $\pi_1(A_1) \neq \{1\}$, then (A, i) admits a covering $p : (B, j) \to (A, i)$ satisfying (BD), (FV), and (Inf).

8.5　**Example.** Suppose that (A, i) has the form

(1)　　　　$(A, i) = $

Since, by (Min), (A, i) has no terminal vertices, we necessarily have

(2)　　　　　　　　　　　　　$u'v' \geq 1.$

We assume further that

(3)　　　　　　　　　　　　　$m > 1.$

Let

(4)　　$(B, j) = $

where

(5)　　$(E, j) = $

A little reflection shows that there is a covering

(6)　　　　　　　　　　$p : (B, j) \to (A, i)$

compatible with the notation, and (B, j) is clearly unimodular and infinite. In (E, j) we calculate

$$\frac{\Delta e_1}{\Delta e_0} = \frac{\Delta d_1}{\Delta d_0} \cdot \frac{v'}{t} \cdot \frac{t}{1} \frac{\Delta d_0}{\Delta d_1} \cdot \frac{s}{1} \cdot \frac{\Delta c_0}{\Delta c_1} \frac{u'}{r} \cdot \frac{r}{1} \cdot \frac{\Delta c_1}{\Delta c_0} = u'v's$$

Hence, if e_0 and e_0' are two consecutive copies of e_0 in (B, j), then in (B, j) we have

(7)　　　　　　　　　$\frac{\Delta e_0'}{\Delta e_0} = (u'v's) \cdot \frac{m}{s} = u'v'm.$

It follows that, if a_0 is the initial vertex of (B, j), b is a vertex of (E, j), and b_n is the incarnation of b in the nth copy of (E, j) in (B, j), then, using (7),

(8)
$$\frac{\Delta b_{n+1}}{\Delta a_0} = \frac{r}{u} \cdot \frac{\Delta c_1}{\Delta c_0} \cdot \frac{m}{s} \cdot (u'v'm)^n \cdot \frac{\Delta b}{\Delta e_0}.$$

Since E is finite, it follows from (2) (3), and (8) that (B, j) has bounded denominators and finite volume.

Conclusion

(9)   If (A, i) has the form (1) above with $m > 1$, then (A, i) admits a covering $p : (B, j) \rightarrow (A, i)$ satisfying (BD), (FV), and (Inf).

(10)   The same conclusion applies also (with simpler constructions and arguments) when (A, i) has the simpler form, with $m > 1$,

In fact, (A, i) is covered by

which has the form of (8.5)(1).

(11)   The same conclusion applies also when (A, i) has the form, with $m > 1$,

In fact, if C has a terminal vertex $\neq c_1$ then, in view of (Min), (A, i) has the form of (8.5)(1). This and (10) covers the case when C is a tree. Otherwise $\pi_1(C) \neq \{1\}$ and C admits a double cover $q : C' \rightarrow C$, which we index so that q is index preserving. Let $q^{-1}(c_1) = \{b_0, b_1\}$. Then (A, i) is covered by

which has the form of (8.5)(1).

8.6   Proof of Theorem (8.2)

If A contains a separating edge e such that $i(e)'i(\bar{e})' > 1$, then Example (8.3)(a), (7) applies. Suppose now that A is a tree.

First suppose that, for some $e \epsilon EA$, $i(e) = m_0 \geq 3$. Let e be the first edge of a path to a terminal vertex of (the tree) A, with $\bar{f}$ the last edge of the path. Then $i(f) = m_1 \geq 2$ by condition (Min) ((A, i) has no terminal vertices). Thus (A, i) has the form

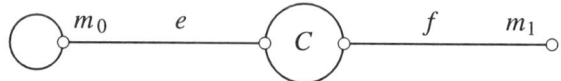

with $m_0 \geq 3$ and $m_1 \geq 2$, so Example (8.3)(b)(7) applies (or else, in case $f = \bar{e}$, we have case (8.3)(a) again.)

Suppose, finally, that $i(e) \leq 2$ for all $e \epsilon EA$. By condition (NDR) there is an $e \epsilon EA$ with $i(e) = 2$ and $E_0(\partial_0 e) \neq \{e\}$. Now embed e in a path from one terminal vertex of A to another. Then (A, i) has the form

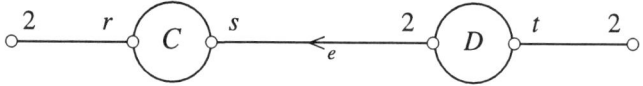

or

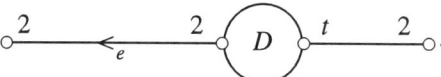

These cases are covered by Examples (8.5)(9) and (10), respectively.

8.7   Remarks

1. Suppose that (A, i) has the form

$$(A, i) = \left((A_0, i) \;\bigcirc_a\; (A_1, i) \right)$$

and that we have succeeded in constructing a covering

$$p_0 : (B_0, j) \to (A_0, i)$$

satisfying (BD), (FV), and (Inf). Then we can obtain a covering

$$p : (B, j) \to (A, i)$$

with the same properties simply by attaching a copy of (A_1, i) to (B_0, i) at each vertex in $p_0^{-1}(a)$. The same remark applies if (A, i) is obtained from (A_0, i) by attaching several pairwise disjoint rooted graphs at various vertices of A_0.

2. The *barycentric subdivision* (A', i') of an edge-indexed graph (A, i) is obtained by replacing each

$$a \underset{e}{\overset{\displaystyle i(e) \qquad\qquad\qquad i(\bar{e})}{\text{———————————}}} b \quad \text{in} \quad (A, i)$$

by

$$\underset{i'(e_0) = i(e)}{\overset{e_0}{\longrightarrow}} \underset{1}{\overset{m(e) = m(\bar{e})}{}} \underset{1}{\overset{}{}} \underset{i(\bar{e}) = i'(\bar{e}_1)}{\overset{e_1}{\longrightarrow}} \quad \text{in} \quad (A', i'),$$

where $m(e)(= m(\bar{e}))$ denotes the barycenter vertex of e, and we have $i'(e_0) = i(e)$, $i'(\bar{e}_0) = 1 = i'(e_1)$, $i'(\bar{e}_1) = i(\bar{e})$.

Properties like (U), (BD), (FV), (Inf), … for (A, i) are clearly inherited by (A', i'). A covering $p : (B, j) \to (A, i)$ defines in the obvious way, a covering $p' : (B', j') \to (A', i')$.

We next treat some cases which have no separating edges, as are present in all the cases treated above.

8.8 Examples. Loops and cages

For integers $r, s, m > 0$ we put

$$L_r^m = \text{a bouquet of } m \text{ unimodular loops of index } r,$$

(1)
$$L_r^1 = \overset{r}{\underset{r}{\bigcirc}} \;, \qquad L_r^2 = \overset{r \qquad\qquad r}{\underset{r \qquad\qquad r}{\infty}}$$

and

(2)
$$C_{r,s}^m = c_0 \overset{r}{\underset{r}{\bigg/}} \overset{\overset{\vdots}{\uparrow}}{m} \overset{\vdots}{\underset{\downarrow}{}} \overset{s}{\underset{s}{\bigg\backslash}} c_1.$$

The latter denotes the *cage*, with vertices c_0, c_1, $E_0(c_0) = \{e_1, \ldots, e_m\}$, and, for $h = 1, 2, \ldots, m$, $\partial_1(e_h) = c_1$, $EC_{r,s}^m = E_0(c_0) \cup \overline{E_0(c_0)}$, $i(e_h) = r$, and $i(\bar{e}_h) = s$. Clearly L_r^m and $C_{r,s}^m$ are unimodular.

For typographical convenience we shall use the abbreviation,

$$C_{r,s}^m = \circ\!\!\underline{C_{r,s}^m}\!\!\circ\,.$$

Notice that there are double covers

(3) $$C_{r,r}^m \to L_r^m$$

(collapsing the two vertices, but no edge), and

$$p : C_{r,r}^2 \to L_r^1,$$

(4)

$$p(\bar{e}) = g = p(f).$$

(collasping both vertices and edges). Thus, to obtain coverings of $L_r^m (r > 1, \, m \geq 1)$ of the type we seek, it suffices to do so for $C_{r,r}^m$ $(r, m > 1)$. We shall obtain these below.

Suppose that $C_{r,s}^m$ satisfies (NDR) and (Min). If $m = 1$, then $C_{r,s}^1 = \circ\!\!\underline{rs}\!\!\circ$ and $r, s \geq 2$, by (Min), and r or $s \geq 3$, by (NDR). This case is covered by (8.3)(7). So assume henceforth that

(5) $$m > 1.$$

Then (Min) is automatic, and (NDR) $\Leftrightarrow rs > 1$. There are then three possibilities, which we treat separately: (i) $r's' > 1$; (ii) $r = s = 2$; (iii) $s = 1$.

(6) If $r's' > 0$ then there is a covering $p : B_{r,s}^m \to C_{r,s}^m$, where

$$B_{r,s}^m = \underset{a_0}{\circ}\!\!\underline{C_{r,s'}^m}\!\!\underset{a_1}{\circ}\!\!\underline{C_{1,r'}^m}\!\!\underset{a_2}{\circ}\!\!\underline{C_{1,s'}^m}\!\!\underset{a_3}{\circ}\!\!\underline{C_{1,r'}^m}\!\!\underset{a_4}{\circ}\!\!\cdots\cdots$$

Clearly $B_{r,s}^m$ is unimodular and infinite. In $B_{r,s}^m$ we have

(7)
$$\frac{\Delta a_{2n}}{\Delta a_0} = \frac{s'}{r} \cdot \frac{r'}{1} \cdot \frac{s'}{1} \cdots = \frac{(r's')^n}{r}, \quad \text{and}$$

$$\frac{\Delta a_{2n+1}}{\Delta a_0} = \frac{\Delta a_{2n}}{\Delta a_0} \cdot s'.$$

From (7) it follows that $B_{r,s}^m$ has bounded denominators, and clearly

(8) $$\mathrm{Vol}(B_{r,s}^m) < \infty \quad \text{iff } r's' > 1.$$

Conclusion

(9)   If $r's' > 1$ then the covering $p : B_{r,s}^m \to C_{r,s}^m$ satisfies (BD), (FV), and (Inf).

Next consider

(10)
$$C(r, s) = c_0 \overset{r}{\underset{s}{\bigcirc}} \overset{r}{} c_1, \quad \text{with } r, s > 1.$$

This admits the covering

$$p : D(r, s) \to C(r, s), \quad \text{where}$$

(11) $D(r, s) =$

$$p(a_n) = c_{(n \bmod 2)}, \qquad p(b_n) = c_{(n' \bmod 2)}.$$

Clearly $D(r, s)$ is unimodular, and

$$\frac{\Delta a_{2n}}{\Delta a_0} = (rs)^n = \frac{\Delta b_{2n}}{\Delta a_0}, \quad and$$

$$\frac{\Delta a_{2n+1}}{\Delta a_0} = (rs)^n r = \frac{\Delta b_{2n+1}}{\Delta a_0}.$$

Thus $D(r, s)$ has (BD) and (FV), since $rs > 1$.

More generally, consider also

(12)
$$C(r, s, A) = c_0 \overset{r}{\underset{s}{\bigcirc}}\!\!\xrightarrow{\ A\ }\!\! \overset{r}{\underset{s}{}} c_1,$$

where $r, s > 1$ and A is a finite unimodular edge-indexed graph with $\frac{\Delta c_1}{\Delta c_0} = 1$ in A.

Let

(13) $D(r, s, A)$ be obtained from $D(r, s)$ by attaching, for each $n \geq 0$, a copy of A by identifying $p(a_n)$ (resp., $p(b_n)$) in A with a_n (resp., b_n) in $D(r, s)$.

For n even, we have in $D(r, s, A)$:

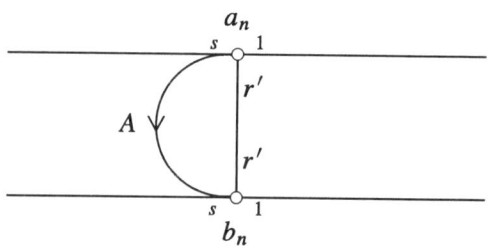

For n odd, r and s are exchanged, and the orientation of A is reversed. Then the covering p above obviously extends to a covering $p: D(r, s, A) \to C(r, s, A)$. Moreover $D(r, s, A)$ clearly preserves the properties (U), (BD), and (FV) of $D(r, s)$.

Conclusion

(14) $C(r, s, A)$ as in (12) admits the covering $p : D(r, s, A) \to C(r, s, A)$ with $D(r, s, A)$ satisfying (BD), (FV), and (Inf).

   Note that $C_{2,2}^2 = C(2, 2)$, and, for $m > 2$, $C_{2,2}^m$ is of the form $C(2, 2, A)$.

(15)   **Corollary.** *Suppose that (A, i) contains (unimodular) loop $L_r \; (= L_r^1) =$*
$$a \underset{r}{\overset{r}{\bigcirc}}$$
with $r > 1$. Then there is a covering $p : (B, j) \to (A, i)$ satisfying (BD), (FV), *and* (Inf).

Proof. Writing $(A, i) = \left(L_r \underset{}{\overset{}{\bigstar}} A_1\right)$, and applying (8.7), Remark 1, we reduce to the case $(A, i) = L_r$. By (4), L_r is covered by $C_{r,r}^2$. If $r \geq 3$, then (9) applies to give the result. If $r = 2$, then (10) and (11) give the desired covering.

   Among cages, it remains to consider

(16)
$$\underset{C_{r,1}^m}{\circ\!\!-\!\!\!-\!\!\!-\!\!\!-\!\!\circ} = \overset{r}{\underset{r}{\bigcirc}}\!\!\overset{\uparrow}{\underset{\downarrow}{\,m\,}}\!\!\overset{1}{\underset{1}{}}, \quad m > 1, \; r > 1.$$

Here we construct a covering supplied to us by Lisa Carbone.

   First observe that $C_{r,1}^m$ is double covered by

(17)
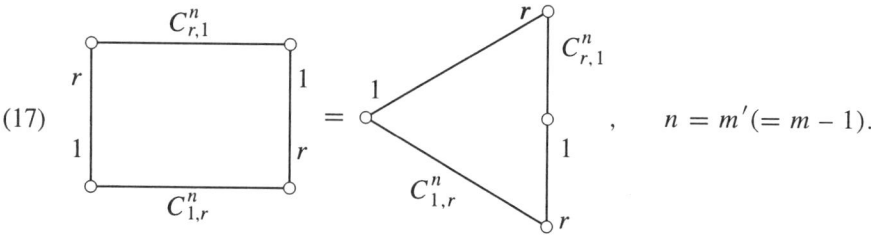
$, \qquad n = m'(= m - 1)$.

Writing

(18) $r = u + v, \quad u, v > 0,$

we can then cover (17) by

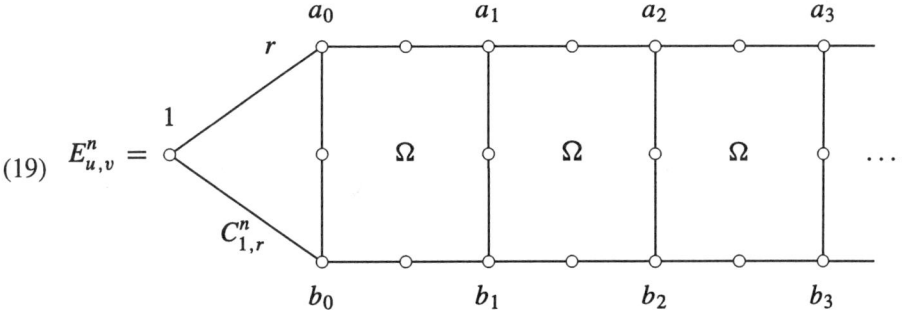

(19) $E^n_{u,v} =$

where

(20) $\Omega =$

It is readily seen that $E^n_{u,v}$ is unimodular, and

$$\frac{\triangle a_n}{\triangle a_0} = \left(\frac{r}{v}\right)^n = \frac{\triangle b_n}{\triangle a_0} \quad \text{for } n \geq 0.$$

It follows easily that, since $\frac{r}{v} > 1$ (cf. (18)), we have (FV). Moreover we have (BD) provided that $v | r$. This can be achieved by taking $v = 1$.

Conclusion

(21)   For $m, r > 1$, there is a covering $p : E^{m'}_{r',1} \to C^m_{r,1}$, where $E^{m'}_{r',1}$ (as in (19), (20)) satisfies (BD), (FV), and (Inf).

Finally, we shall have need of the case

(22) $E(r, s, C) =$

where $r > 1$, $s \geq 1$, and C is a finite unimodular edge-indexed graph in which $\frac{\Delta_\circ}{\Delta_\bullet} = 1$. Then:

(23)   There is a covering $p : F(r, s, C) \to E(r, s, C)$, where

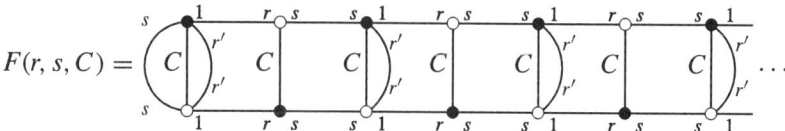

satisfies (BD), (FV), and (Inf).

The covering p is essentially explained by the notation, and the indicated properties of $F(r, s, C)$ are readily verified.

(24)   In the absence of C ($C = \emptyset$), we similarly obtain the desired kind of covering

of $\bigcirc$, by suppressing all appearances of C in $F(r, s, C)$.

8.9 Two vertex graphs

We shall here verify Theorem (8.1)(2) for all two vertex graphs. When $VA = \{a_0, a_1\}$, (A, i) has the form

(1) $(A, i) = \left(A_0 \;\; C \;\; A_1 \right)$,

where (A_0, i) and (A_1, i) are bouquets of unimodular loops $L_r (= L_r^1) = \bigcirc$, and (C, i) is a unimodular *cage*

(2) $(C, i) = \begin{array}{c} e_1 \\ p_1 \quad q_1 \\ \vdots \\ p_\ell \quad q_\ell \\ e_\ell \end{array}$.

If (A_0, i) or (A_1, i) contains a loop, L_r with $r > 1$, then we can find a covering $p : (B, j) \to (A, i)$ satisfying (BD), (FV), and (Inf), thanks to (8.8)(15).

Suppose, on the other hand, that all loops L_r in (A_0, i) and (A_1, i) are unramified ($r = 1$). Then condition (NDR) implies that some p_h or q_h is > 1. We are at liberty to delete loops if we wish; they can then be reattached after constructing the desired covering

First consider the case $\ell = 1$:

$$(A, i) = \boxed{A_0} \; \underset{}{\overset{p \qquad\qquad q}{\circ\!\!-\!\!-\!\!-\!\!-\!\!-\!\!-\!\!-\!\!-\!\!-\!\!-\!\!-\!\!\circ}} \; \boxed{A_1}, \quad \text{say } p \geq q.$$

If $p'q' > 1$ this is a case of example (8.3)(7). Otherwise $(p, q) = (2, 2)$ or $(2,1)$. Then conditions (NDR) and (Min) imply that (A, i), with perhaps some loops deleted, has the form

$$(D, i) = \bigcirc\!\!-\!\!\overset{2}{\!-\!-\!-\!-\!}-\!\!\overset{2}{\!-\!-\!-\!}\!\!\circ, \quad \text{or}$$

$$(E, i) = \bigcirc\!\!-\!\!\overset{2}{\!-\!-\!-\!-\!}-\!\!\overset{1}{\!-\!-\!-\!}\!\!\bigcirc \, .$$

(Recall that all loops are unramified.) Now (D, i) is covered by

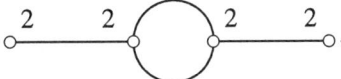

which is a case of Example (8.5)(10). In turn, (E, i) is covered by

$$(E_1, i) = \bigcirc\!\!\overset{2}{-\!-\!}\!\!\overset{1}{-\!-\!}\!\!\bigcirc\!\!\overset{1}{-\!-\!}\!\!\overset{2}{-\!-\!}\!\!\bigcirc,$$

which itself is covered by

$$(E_2, i) = \bigcirc\!\!\overset{2}{-}\!\overset{1}{-}\bigcirc\!\!\overset{1}{-}\!\!\underset{u}{\overset{2}{-}}\bigcirc\!\!\overset{2}{-}\!\overset{1}{-}\bigcirc\!\!\overset{1}{-}\!\!\underset{v}{\overset{2}{-}}\bigcirc .$$

Removing the left and right loops, and grouping terms between u and v yields a case of Example (8.5)(1). Thus, using Remark (8.7)(1), we can, in each of the above cases, produce a covering of (A, i) satisfying (BD), (FV), and (Inf).

Now consider the case $\ell > 1$. Unimodularity of (A, i) means that p_h/q_h is independent of $h = 1, 2, \dots, \ell$. Writing this fraction in reduced form

$$p_h/q_h = r/s, \qquad gcd(r, s) = 1,$$

we have, for each h,

$$p_h = m_h r, \qquad q_h = m_h s, \quad \text{for some integer } m_h > 0.$$

Thus, removing the loops from (A, i), we have

(3)
$$(C, i) = \quad \overset{m_1 r}{\underset{m_\ell r}{\Big(\ \vdots\ \Big)}}\overset{m_1 s}{\underset{m_\ell s}{}} .$$

Clearly (C, i) is covered by

(4)
$$C^m_{r,s} = \overset{r\ \uparrow\ s}{\underset{r\ \downarrow\ s}{\Big(\ m\ \Big)}} ,$$

where $m = m_1 + \cdots + m_\ell > 1$, $gcd(r, s) = 1$, and, say, $r \geq s$. If $r's' > 1$, then we can find the desired covering of $C^m_{r,s}$, using Example (8.8)(9). Otherwise $(r, s) = (r, 1)$, $r \geq 1$. If $r \geq 2$, then Example (8.8)(21) gives the desired covering.

Suppose that $r = 1 (= s)$. This means that $p_h = q_h = m_h$ for all h. Recall that $p_h > 1$ for at least one h, by (NDR); say $p_1 > 1$. Moreover

$$m := m_1 + \cdots + m_\ell = p_1 + \cdots + p_\ell \geq 3$$

since otherwise $\ell = 1$, the case already treated. It is easily seen then that we can cover (C, i) by

$$E(2, 1, D) = \overset{1}{\underset{1}{\Big(\ D\ \Big)}}\overset{2}{\underset{2}{}} \quad \text{(as in (8.8)(22)),}$$

where $D = C^n_{1,1}$, $\quad n = m - 3\ (\geq 0)$.
(In case $n = 0$, D is suppressed.)

The desired covering in this case is furnished by (8.8)(23) and (24).

Chapter 9

Parabolic Actions, Lattices, and Trees

9.0 Introduction

Let ε be an end of a tree X. Then VX partitions into "horospheres" X_n $(n\epsilon\mathbf{Z})$ so that if $e\epsilon EX$ is directed toward ε, and $\partial_0 e\epsilon X_n$, then $\partial_1 e\epsilon X_{n+1}$. Such edges thus define maps

(1)
$$\cdots \to X_{n-1} \to X_n \to X_{n+1} \to \cdots, \quad \text{with}$$
$$\lim_{\longrightarrow} X_n = \{\varepsilon\} \quad \text{and} \quad \lim_{\longleftarrow} X_n = Ends(X) - \{\varepsilon\}.$$

Moreover, (1) completely determines (X, ε). As in (4,9)(10)–(11), we call X a *parabolic tree* if it has a unique end.

Suppose that $\Gamma \leq G = \text{Aut}(X)$ fixes ε. Then there is an exact sequence

$$1 \to \Gamma^0 \to \Gamma \xrightarrow{\tau} \mathbf{Z}$$

defined by $g(X_n) = X_{n+\tau(g)}$, for all n. If Γ is discrete, we have $\Gamma = \Gamma^0$. We call a tree *action parabolic* if Γ fixes a unique end, but no vertex, of X. We show (Theorem (9.7)) that if a locally finite tree X admits a parabolic lattice Γ, then $\ell(\Gamma) = 0$, every X-lattice is parabolic, and X is a parabolic tree.

The remainder of Chapter 9 is devoted to a study of parabolic tree lattices, mainly through examples. We first study "parabolic ray lattices," i.e., those parabolic lattices for which

$$\Gamma\backslash X = \circ\!\!-\!\!-\!\!-\!\!\circ\!\!-\!\!-\!\!-\!\!\circ\!\!-\!\!-\!\!-\!\!\circ\!\!-\!\!-\!\!-\!\!\circ \cdots$$

This situation is equivalent to presenting Γ as the union of a sequence

$$\Gamma_0 \leq \Gamma_1 \leq \Gamma_2 \leq \cdots$$

of finite subgroups $\Gamma_n \neq \Gamma$, such that Γ_0 contains no non-trivial normal subgroup of Γ, and $\sum_{n\geq 0} |\Gamma_n|^{-1} < \infty$. We provide then, in (9.8), a general procedure for calculating

$$Z_G(\Gamma) \cong \left(\bigcap_{n\geq 0} N_\Gamma(\Gamma_n)\right) / \Gamma_0,$$

$$N_G(\Gamma) \cong \Gamma \rtimes_{\Gamma_0 = ad_\Gamma(\Gamma_0)} \mathrm{Aut}^{\mathrm{filt}}(\Gamma),$$

and

$$N_G(\Gamma) / \Gamma \cong \mathrm{Aut}^{\mathrm{filt}}(\Gamma) / ad_\Gamma(\Gamma_0).$$

Several examples are computed, showing that $Z_G(\Gamma)$ and $N_G(\Gamma)/\Gamma$ can be either infinite or finite, and even trivial.

A notable example (9.10), is when $\Gamma_n = S_n$, the symmetric group, so $\Gamma = S_\infty$, which contains the (simple) alternating group $\Gamma' = A_\infty$ with index 2. In this case we calculate, for example, that

$$Z_G(\Gamma) \cong S_2, \quad Z_G(\Gamma') \cong S_3$$
$$N_G(\Gamma) \cong \Gamma \times S_2, \quad N := N_G(\Gamma') \cong \Gamma \times S_3, \quad \text{and}$$
$$N = C_G(\Gamma) = C_G(\Gamma') = C_G(N).$$

Thus N furnishes *a (parabolic) tree lattice that is its own commensurator*. However the tree X here is of unbounded degree.

In (9.16) we show that tree lattices can be simple groups only when they are parabolic.

To produce some of these phenomena on a tree of bounded degree, we make use of an ingenious group theoretic construction of Peter Neumaun, kindly supplied to us on request. His Example (9.12) furnishes a parabolic lattice Γ on an X of bounded degree such that $Z_G(\Gamma) \cong \Gamma_0$ and $N_G(\Gamma)/\Gamma \cong \Gamma_1$ are finite groups. We have not succeeded in determining, in this case, whether $C_G(\Gamma)/\Gamma$ is finite.

The above examples concern the case when $A = \Gamma \backslash X$ is the simplest possible parabolic tree, a ray. In the last sections we explore how complicated A can be. In particular we produce ((9.14)) (somewhat surprising) examples for which the horospheres A_n of A are infinite for all $n \in \mathbf{Z}$, and even ((9.15)) such examples that have bounded degree.

9.1 Ends(X)

We review some properties of ends of trees (cf. [AB], (2.23), (2.24), (2.33), and [B2], §1). Let X be a tree. An *X-ray (from $x_0 \in VX$)* is an infinite reduced path $\gamma = (e_1, e_2, e_3, \dots)$ in X with $\partial_0 e_1 = x_0$. It is determined by its vertex sequence

$x_0, x_1, x_2, x_3, \ldots$, where $x_n = \partial_1 e_n = \partial_0 e_{n+1}$ for $n > 0$. We shall sometimes inden-
tify γ with its underlying (linear) subtree of X. Call two X-rays γ and γ' equivalent,
denoted $\gamma \sim \gamma'$, if $\gamma \cap \gamma'$ is an X-ray. The equivalence classes form the space
$Ends(X)$ of ends of X.

Let $\varepsilon \in Ends(X)$. For $x \in VX$ there is a unique X-ray, denoted $[x, \epsilon)$, from x "to
ε," i.e., in the class of ε. For $x, y \in VX$,

(1)

$$[x, \varepsilon) \cap [y, \varepsilon) = [u, \epsilon) \quad \text{for some} \quad u \in VX.$$

For $v \in [x, \varepsilon) \cap [y, \varepsilon)$,

(2) $$(x - y)_\varepsilon := d(y, v) - d(x, v)$$

is clearly independent of the choice of v. Moreover,

(3) $$(x - y)_\varepsilon + (y - z)_\varepsilon = (x - z)_\varepsilon \quad \forall\, x, y, z \in VX.$$

The group $G = \mathrm{Aut}(X)$ acts on $Ends(X)$, and, by transport of structure, we have

(4) $$(gx - gy)_{g\varepsilon} = (x - y)_\varepsilon \quad \forall\ g \in G.$$

If $\varepsilon' \neq \varepsilon$ is another end of X, then there is a unique bi-infinite linear tree

(5) $$(\varepsilon', \varepsilon) \subset X$$

defined by

(6) $$V(\varepsilon', \varepsilon) = \{x \in VX \mid [x, \varepsilon) \cap [x, \varepsilon') = \{x\}\}.$$

Further,

(7) $$(\varepsilon', \varepsilon) = [x, \varepsilon) \cup [x, \varepsilon') \quad \forall\ x \in V(\varepsilon', \varepsilon).$$

9.2   Horospheres and horoballs

Let $\varepsilon \in Ends(X)$ and choose a base point $x_0 \in VX$. For $r \in \mathbf{Z}$ put

(1)
$$X_r = \{x \in VX \mid (x - x_0)_\varepsilon = r\}$$
$$VB_r = \{x \in VX \mid (x - x_0)_\varepsilon \geq r\}.$$

Then VB_r is the vertex set of a subtree B_r of X, called a *horoball centered at ε*. We
call each X_r a *horosphere centered at ε*. Notice that changing the base point x_0 has
only the effect of translating the indices on X_r and B_r.

Clearly

(2) $$V B_r = X_r \amalg V B_{r+1}, \quad \text{so} \quad B_r \supset B_{r+1}.$$

Further there is a canonical map

(3) $$q : X_r \to X_{r+1} \quad \text{defined by}$$
$$\{q(x)\} = [x, \varepsilon) \cap X_{r+1}.$$

In this way, the *ended tree* (X, ε) defines a direct (bi-infinite) sequence of sets

(4) $$(X_r, \; q : X_r \to X_{r+1})_{r \in \mathbf{Z}}, \quad \text{with}$$

(5) $$\lim_{\overrightarrow{r}} X_r = \{\varepsilon\}, \quad \text{a one point set.}$$

Conversely, the system (4) satisfying (5) determines (X, ε), in the following sense. Given any direct sequence of sets, $q : X_r \to X_{r+1}$ $(r \in \mathbf{Z})$, with $\lim_r X_r = \{\varepsilon\}$, a one point set, we define a simplicial graph X with $VX = \amalg_{r \in \mathbf{Z}} X_r$ and with oriented edges $e_x = (x, q(x))$ and $\bar{e}_x = (q(x), x)$ $\forall x \in VX$. This graph manifestly has no circuits (it is a forest), and it is connected thanks to condition (5); thus X is a tree. For all $x \in VX$, the vertex paths $(x, q(x), q^2(x), q^3(x), \dots)$ are X-rays all leading to the same end, ε, of X. Further, it is clear that

(6) $$\lim_{\overleftarrow{r}} X_r = Ends(X) - \{\varepsilon\}.$$

9.3   End stabilizers

Let $G = \text{Aut}(X)$, $\varepsilon \in Ends(X)$, and

(1) $$G_\varepsilon = \{g \in G \mid g\varepsilon = \varepsilon\}.$$

Inversions fix no ends, so G_ε contains none. For $x, y \in VX$ and $g \in G_\varepsilon$ we have

$$
\begin{aligned}
(gx - x)_\varepsilon - (gy - y)_\varepsilon &= (gx - gy)_\varepsilon - (x - y)_\varepsilon && ((9.1)(3)) \\
&= (gx - gy)_{g\varepsilon} - (x - y)_\varepsilon && (g\varepsilon = \varepsilon) \\
&= (x - y)_\varepsilon - (x - y)_\varepsilon && ((9.1)(4)) \\
&= 0.
\end{aligned}
$$

Thus we can define

(2) $$\tau_\varepsilon : G_\varepsilon \to \mathbf{Z}, \quad \tau_\varepsilon(g) := (gx - x)_\varepsilon \qquad \forall x \in VX.$$

For $g, h \in G_\varepsilon$, $\tau_\varepsilon(gh) = (ghx - x)_\varepsilon = (ghx - hx)_\varepsilon + (hx - x)_\varepsilon = \tau_\varepsilon(g) + \tau_\varepsilon(h)$, so

(3) τ_ε is a homomorphism;   put $G_\varepsilon^0 = Ker(\tau_\varepsilon)$.

Since an inversion fixes no ends of X,

(4) G_ε contains no inversions.

Taking $x \in X_g$ in (2) we see that

(5) $\ell(g) = |\tau_\varepsilon(g)| \quad \forall\, g \in G_\varepsilon.$

9.4   Parabolic actions

Let $\Gamma \leq G = Aut(X)$ fix the end ε of X : $\Gamma \leq G_\varepsilon$. Choose $x_0 \in VX$ and let $x_0, x_1, x_2, \ldots$ be the vertex sequence of the ray $[x_0, \varepsilon)$. In the exact sequence

(1) $1 \to \Gamma^0 \to \Gamma \xrightarrow{\tau_\varepsilon} \mathbf{Z}$

             $(\Gamma^0 = Ker(\tau_\varepsilon)$; $\ell(g) = |\tau_\varepsilon(g)|$ for $g \in \Gamma)$,

we have $((5.2)(c))$

(2) $\Gamma_{x_n} = \Gamma_{x_n}^0 \leq \Gamma_{x_{n+1}} \quad \forall\, n \geq 0$ and $\Gamma^0 = \bigcup_{n \geq 0} \Gamma_{x_n}.$

Suppose that $g \in \Gamma$ and $\tau_\varepsilon(g) = t > 0$. Then ε must be one of the two ends (of X_g) fixed by g, so $x_n \in X_g$ for $n \gg 0$. Thus,

(3)   For $n \gg 0$, $gx_n = x_{n+t}$, so $\Gamma_{x_n} \leq \Gamma_{x_{n+t}} = g\Gamma_{x_n} g^{-1}.$

   Suppose that Γ is *discrete*. Then the groups Γ_{x_n} are finite, so in (3), we have $\Gamma_{x_n} = \Gamma_{x_{n+t}}$ for $n \gg 0$, and so $\Gamma^0 = \Gamma_{x_n}$ for $n \gg 0$. Further since g normalizes Γ^0, Γ^0 fixes all points of the g-axis X_g. This applies to all hyperbolic $g \in \Gamma$, so Γ^0 acts trivially on $X_\Gamma = \bigcup_{g \in \Gamma, \ell(g) > 0} X_g$ (see (5.7)). Since $\Gamma/\Gamma^0 (\leq \mathbf{Z})$ is cyclic, we conclude that $X_\Gamma = X_g$ for all hyperbolic $g \in \Gamma$, and Γ acts on X_Γ by translation. Thus:

(4)   If Γ is discrete and $\ell(\Gamma) \neq 0$, then X_Γ is a linear tree on which Γ acts by translation, fixing its two ends. Further Γ^0 is finite and acts trivially on X_Γ.

   Returning to the general situation,

(5)   The Γ-action on X is called a *parabolic action* if
          (a) Γ fixes a unique end, ε, of X, and
          (b) Γ fixes no vertex of X.

From (4) and (5) (a) we conclude that:

(6)   If Γ is discrete and acts parabolically on X, then $\Gamma = \Gamma^0$, i.e., $\ell(\Gamma) = 0$.

(7)   Let Γ fix ε, as above. The Γ- action on X is parabolic iff the stabilizers Γ_{x_n}
         strictly increase infinitely often, iff Γ^0 fixes no vertices of X.

To see this suppose first that Γ acts parabolically and $\Gamma^0 = \Gamma_{x_n}$ for some n. If
$\Gamma = \Gamma^0$ this contradicts assumption (5)(b). If $\Gamma \neq \Gamma^0$, i.e., $\ell(\Gamma) \neq 0$ then, since
$\Gamma^0 \triangleleft \Gamma$, the tree of fixed points of Γ^0 is Γ-invariant, hence Γ^0 acts trivially on X_Γ.
Since Γ'/Γ^0 is cyclic, X_Γ must be a linear tree on which Γ acts by translation, fixing
its two ends, contrary to assumption (5)(a).

Suppose, conversely, that $\Gamma^0 \neq \Gamma_{x_n}$ $\forall n \geq 0$. Since Γ fixes ε, it follows easily
that Γ^0, hence also Γ, has no fixed points, whence (5)(b). If Γ fixes an end $\varepsilon' \neq \varepsilon$,
then Γ leaves invariant the axis $(\varepsilon', \varepsilon)$, on which it must act by translation, with Γ^0
acting trivially hence with fixed points, contrary to our first observation above. This
proves (7).

Consider the horospheres, X_r, and horoballs, B_r $(r \in \mathbf{Z})$, centered at ε; the
indexing, depending on a choice of base point $x_0 \in VX$, is unique up to translation.
We have

(8) $h X_r = X_{r+\tau_\varepsilon(h)}, \quad h B_r = B_{r+\tau_\varepsilon(h)} \quad \forall h \in \Gamma.$

(9) Γ^0 stabilizes each X_r and B_r; hence Γ^0 does not act minimally on X.

Further, the sequence $q : X_r \to X_{r+1}$ $(r \in \mathbf{Z})$ defining (X, ε) is equivariant for the
Γ^0-action, so, by passage to the Γ^0-quotient, we obtain a system $q : A_r \to A_{r+1}$ $(r \in$
$\mathbf{Z})$, $A_r = \Gamma^0 \backslash X_r$, defining an ended tree, (A, ε_A), $A = \Gamma^0 \backslash X$. Thus we have the
ended tree quotient

(10) $p : (X, \varepsilon) \longrightarrow (A, \varepsilon_A) = \Gamma^0 \backslash (X, \varepsilon).$

In particular, for the edge-indexed quotient

$$(A, i) = I(\Gamma^0 \backslash\backslash X),$$

the fact that Γ stabilizers in X increase as one moves toward ε translates into the
condition that

(11) $i(e) = 1 \quad \forall e \in EA$ directed toward ε_A.

In general, when a group H acts on a tree X, we call a *subgroup* $P \leq H$ *parabolic*
(on X) if the action of P on X is parabolic. Let ε_P denote the (unique) end of X fixed
by P. These ends ε_P $(P \leq H$ parabolic on $X)$ will be called the *H-parabolic ends*

of X. For $h \in H$, $h\varepsilon_P = \varepsilon_{hPh^{-1}}$, so the set of them is H-invariant. The *maximal parabolic subgroups* of H are the stabilizers H_ε of H-parabolic ends ε, and every parabolic subgroup is contained in such an H_ε. The H_ε's are all H-conjugate iff H acts transitively in its set of H-parabolic ends of X.

When $H \leq G = \mathrm{Aut}(X)$ (the action is faithful), the set of H-parabolic ends is invariant under the commensurator $C_G(H)$.

9.5   Parabolic trees

Call a tree X a *parabolic tree* if it has a unique end ε. According to (9.2), such a tree is defined by the sequence

$$(1) \qquad\qquad (X_r, \ q : X_r \ \to \ X_{r+1})_{r \in \mathbb{Z}}$$

of its horospheres, satisfying

$$(2) \qquad\qquad \varinjlim_r X_r = \{\varepsilon\}, \quad \text{and}$$

$$(3) \qquad\qquad \varprojlim_r X_r = \emptyset.$$

For $G = \mathrm{Aut}(X)$ we have

$$(4) \qquad\qquad \ell(G) = 0,$$

since a hyperbolic axis has two ends.

9.6   Parabolic lattices

Let X be a locally finite tree and $\Gamma \leq G = \mathrm{Aut}(X)$ an X-lattice. Suppose that the Γ-action on X is parabolic, fixing the end ε. Then
(a) X is a parabolic tree: $Ends(X) = \{\varepsilon\}$;
(b) $\ell(\Gamma) = 0$, i.e., $\Gamma = \Gamma^0$, in the notation of (9.4); and
(c) For $x_0 \in VX$, Γ is the (infinitely increasing) ascending union of the Γ_x as $x \to \varepsilon$ along $[x_0, \varepsilon)$.
Since Γ is discrete, (b) and (c) follow from (9.4) (6) and (8). It remains to prove (a). If γ is a reduced path in X moving away from ε, then γ injects into $A = \Gamma \backslash X$ (cf. (9.4)(11)), and stabilizers decrease along γ. Since $\mathrm{Vol}(\Gamma \backslash\backslash X) < \infty$, it follows that γ must be finite. This implies (a), because an end $\varepsilon' \neq \varepsilon$ would allow an infinite path along $(\varepsilon, \varepsilon')$ moving away from ε.

9.7   **Theorem.** *Let X be an infinite locally finite tree that admits a lattice.*

(1)   *The following conditions are equivalent.*

    (a) *Some X-lattice Γ is parabolic.*

    (a′) *Every X-lattice Γ is parabolic.*

    (b) $\ell(\Gamma) = 0$ *for some X-lattice Γ.*

    (b′) $\ell(\Gamma) = 0$ *for every X-lattice Γ.*

    (c) $\ell(G) = 0$, *where $G = \mathrm{Aut}(X)$.*

    (d) *X is a parabolic tree.*

(2)   *When X is not parabolic, we have $\ell(\Gamma) \neq 0$ and $X_\Gamma = X_G$ for all X-lattices Γ.*

Proof of (1). Clearly $(c) \Rightarrow (b') \Rightarrow (b)$, and $(b) \Rightarrow (c)$ by (5.11). Clearly $(a') \Rightarrow (a)$, $(a) \Rightarrow (d)$ by (9.6), and $(d) \Rightarrow (c)$ by (9.5)(4). To conclude, it suffices to prove $(b') \Rightarrow (a')$. Say Γ is an X-lattice and $\ell(\Gamma) = 0$. We want to show that Γ fixes no vertex, but fixes a unique end. If Γ fixes a vertex or inverts an edge then, by discreteness, Γ is finite, hence X is finite, contrary to assumption. It follows then from (5.2)(c) that Γ fixes a unique end of X, whence the proposition.

An infinite locally finite tree X that admits a parabolic lattice, and hence satisfies all conditions of (9.7), will be called a *parabolic lattice tree*.

Part (2) follows from (1) together with (5.7) and (5.11).

9.8   Restriction to horoballs

Let (X, ε) be a parabolic tree, with a lattice $\Gamma \leq G = \mathrm{Aut}(X)$. Pick a base point $x_0 \in VX$, and let

(1) $[x_0, \varepsilon) = $

$$x_0 \quad\quad x_1 \quad\quad x_2 \quad\quad x_3$$
$$\circlearrowleft\!\!-\!\!-\!\!-\!\!\circ\!\!-\!\!-\!\!-\!\!\circ\!\!-\!\!-\!\!-\!\!\circ\!\!-\!\!-\!\!-\!\!\cdots$$

Index the horospheres X_n, and horoballs B_n, so that $x_n \in X_n$. These are G-invariant, so we have a restriction exact sequence

(2) $1 \to G(n) \to G \xrightarrow{\text{res}} \mathrm{Aut}(B_n).$

Then

(3) $\Gamma(n) := \Gamma \cap G(n) \triangleleft \Gamma,$

and

(4) $\Gamma(n) \leq \Gamma_{x_n},$

a finite group. Suppose, on the other hand, that we are given $N \leq \Gamma_{x_n}, N \lhd \Gamma$. Let $Y =$ the tree of fixed points of N. Then $[x_n, \varepsilon) \subset Y$ and Y is Γ-invariant, so $\Gamma \cdot [x_n, \varepsilon) \subset Y$. Now if Γ acts transitively on X_m for $m \geq n$ then $\Gamma \cdot [x_n, \varepsilon) = B_n$, whence $N \leq \Gamma(n)$. Thus:

(5)   If Γ acts transitively on X_m for $m \geq n$, then $\Gamma(n)$ is the largest normal sub-
        group of Γ contained in Γ_{x_n}.

9.9   Parabolic lattices with linear quotient

Let

(1)
$$\Gamma = \bigcup_{r \geq 0} \Gamma_r$$

be a group that is the union of an ascending chain

(2)
$$\Gamma_{-1} = \{1\} \leq \Gamma_0 < \Gamma_1 < \Gamma_2 < \Gamma_3 < \cdots$$

of finite subgroups, with

(3) $[\Gamma_r : \Gamma_{r-1}] = q_r \geq 2$, for all but finitely many r, so $|\Gamma_r| = q_0 q_1 \cdots q_r$ for
        $r \geq 0$, and assume that Γ_0 contains no non-trivial normal subgroup of Γ.

Put

(4)
$$X_r = \Gamma / \Gamma_r$$

so that we have the direct sequence of natural projections

(5)
$$\Gamma \to X_0 \xrightarrow{q} X_1 \xrightarrow{q} X_2 \xrightarrow{q} X_3 \to \cdots, \qquad \text{with direct limit}$$

(6)
$$\varinjlim_r X_r = \{\varepsilon\} = \Gamma / \Gamma \quad \text{(by (1))}.$$

As in (9.2), (5) and (6) define a parabolic tree,

(7)
$$(X, \varepsilon), \qquad VX = \coprod_{r \geq 0} X_r.$$

Moreover left multiplication defines a faithful Γ-action on X, fixing ε, and transitive on each X_r. It follows that

(8)
$$\Gamma \backslash\backslash X = \begin{array}{ccccccc} \Gamma_0 & & \Gamma_1 & & \Gamma_2 & & \Gamma_3 \\ \circ\!\!-\!\!-\!\!-\!\!-\!\!\circ\!\!-\!\!-\!\!-\!\!-\!\!\circ\!\!-\!\!-\!\!-\!\!-\!\!\circ & & & & & & \cdots, \end{array}$$

$$I(\Gamma \backslash\backslash X) = \begin{array}{ccccccc} & \Gamma_0 & & \Gamma_1 & & \Gamma_2 & & \Gamma_3 \\ \circ\!\!-\!\!-\!\!-\!\!\circ\!\!-\!\!-\!\!-\!\!\circ\!\!-\!\!-\!\!-\!\!\circ\!\!-\!\!-\!\!-\!\!\circ & & & & & & \cdots, \\ 1 & q_1 & 1 & q_2 & 1 & q_3 & 1 \end{array} \qquad \text{and}$$

(9) $$\mathrm{Vol}(\Gamma\backslash\backslash X) = \sum_{r\geq 0} \frac{1}{q_0 q_1 \cdots q_r}.$$

In view of (3), $q_0 q_1 \cdots q_r \geq 2^{r-r_0}$, for some r_0, so (9) is finite, i.e.,

(10) Γ is a parabolic X-lattice.

We next calculate $Z_G(\Gamma)$ ((16)) and $N_G(\Gamma)$ ((27)), where $G = \mathrm{Aut}(X)$. Put

$$x_r = 1 \cdot \Gamma_r \quad \in \quad X_r = \Gamma/\Gamma_r,$$

so that $x_0, x_1, x_2, \ldots$ is the vertex sequence of $[x_0, \varepsilon)$. For the Γ-set X_r, one has an exact sequence

(11) $1 \longrightarrow \Gamma_r \longrightarrow N_\Gamma(\Gamma_r) \overset{\zeta_r}{\longrightarrow} \mathrm{Aut}_\Gamma(X_r) \longrightarrow 1,$

where

(12) $\zeta_r(g)(sx_r) = sg^{-1}x_r \qquad (g \in N_\Gamma(\Gamma_r),\ s \in \Gamma).$

An element

$$z \in Z_G(\Gamma) = \mathrm{Aut}_\Gamma(X)$$

is defined by a sequence $z = (z_r)_{r\geq 0}$, $z_r \in \mathrm{Aut}_\Gamma(X_r)$, such that the diagrams

(13)
$$\begin{array}{ccc} X_r & \overset{q}{\longrightarrow} & X_{r+1} \\ z_r \downarrow & & \downarrow z_{r+1} \\ X_r & \underset{q}{\longrightarrow} & X_{r+1} \end{array}$$

commute for all $r \geq 0$; here $q(sx_r) = sx_{r+1}$. From (11) we can write $z_r = \zeta_r(g_r)$, with $g_r \in N_\Gamma(\Gamma_r)$, determined modulo Γ_r. From (13) we have, for $s \in \Gamma$,

$$\begin{aligned} sg_{r+1}^{-1}x_{r+1} &= z_{r+1}(sx_{r+1}) = z_{r+1}(qsx_r) \\ &= qz_r(sx_r) = q(sg_r^{-1}x_r) \\ &= sg_r^{-1}x_{r+1}. \end{aligned}$$

Hence $g_r g_{r+1}^{-1} x_{r+1} = x_{r+1}$, i.e.,

(14)
$$g_r \in \Gamma_{r+1} \cdot g_{r+1} \subset N_\Gamma(\Gamma_{r+1}), \quad \text{and}$$
$$g_{r+1} = \gamma_{r+1} g_r, \quad \gamma_{r+1} \in \Gamma_{r+1}, \quad \text{so } \zeta_{r+1}(g_{r+1}) = \zeta_{r+1}(g_r).$$

It follows inductively from (14) that

(15) $g_0 \in \bigcap_{r\geq 0} N_\Gamma(\Gamma_r), \quad \text{and } z_r = \zeta_r(g_0) \text{ for all } r \geq 0.$

Thus we deduce an isomorphism,

(16) $\zeta : \left(\bigcap_{r \geq 0} N_\Gamma(\Gamma_r) \right) / \Gamma_0 \xrightarrow{\cong} Z_G(\Gamma) = \mathrm{Aut}_\Gamma(X) \quad [\zeta(g)(sx_r) = sg^{-1}x_r].$

Note then that

(17)
(a) $\Gamma_{r_0} \leq \bigcap_{r \geq 0} N_\Gamma(\Gamma_r)$ if $\Gamma_i \lhd \Gamma_{r_0}$, e.g., if $\Gamma_i = \{1\}$, for $i < r_0$;

(b) if $N_\Gamma(\Gamma_r)$ is finite for some r, then $Z_G(\Gamma)$ is finite; and

(c) if $\Gamma_r \lhd \Gamma$ for all $r \geq 0$, then $Z_G(\Gamma) \cong \Gamma / \Gamma_0$.

Next consider $N = N_G(\Gamma)$. Since Γ is transitive on X_0, we have

(18) $$G = \Gamma \cdot G_{x_0}, \qquad \Gamma \cap G_{x_0} = \Gamma_0,$$

so

(19) $$N = \Gamma \cdot N_{x_0}, \qquad \Gamma \cap N_{x_0} = \Gamma_0, \quad \text{and} \quad N = \Gamma \rtimes_{\Gamma_0} N_{x_0}.$$

Since $G_{x_0} \leq G_{x_r}$ for all $r \geq 0$, we conclude that

(20) $$N_{x_0} \leq \bigcap_{r \geq 0} N_G(\Gamma_r).$$

We have the homomorphism

(21)
$$ad_\Gamma : \bigcap_{r \geq 0} N_G(\Gamma_r) \longrightarrow \mathrm{Aut}^{\mathrm{filt}}(\Gamma) := \{\alpha \in \mathrm{Aut}(\Gamma) | \, \alpha\Gamma_r = \Gamma_r, \, \forall r\}$$
$$ad_\Gamma(g)(s) = gsg^{-1}, \quad \text{with kernel } Z_G(\Gamma).$$

On the other hand, any $\alpha \in \mathrm{Aut}^{\mathrm{filt}}(\Gamma)$ defines $\alpha_r : X_r \to X_r$ by $\alpha_r(sx_r) = \alpha(s)x_r$. Then $q(\alpha_r(sx_r)) = q(\alpha(s)x_r) = \alpha(s)x_{r+1} = \alpha_{r+1}(sx_{r+1}) = \alpha_{r+1}(q(sx_r))$. Thus $(\alpha_r)_{r \geq 0}$ defines an automorphism $\rho(\alpha) \in G = \mathrm{Aut}(X)$.

(22)
$$\rho : \mathrm{Aut}^{\mathrm{filt}}(\Gamma) \longrightarrow G = \mathrm{Aut}(X),$$
$$\rho(\alpha)(sx_r) = \alpha(s)x_r \quad \text{for } s \in \Gamma, \, r \geq 0.$$

Clearly ρ is a homomorphism. Taking $s = 1$ in (22) we see that $\rho(\alpha)$ fixes all x_r:

(23) $$Im(\rho) \leq G_{x_0}.$$

Further, for $s \in \Gamma, \alpha(s)x_r = \rho(\alpha)(sx_r) = \rho(\alpha)s\rho(\alpha)^{-1}x_r$, by (23). Taking $r = 0$ we see that $\rho(\alpha)s\rho(\alpha)^{-1} = \alpha(s)\delta(s)$ with $\delta(s) \in G_{x_0}$. Since $s \mapsto \alpha(s)\,\delta(s)$ is a homomorphism $\Gamma \to G$ we have $\delta(st) = \alpha(t)^{-1}\delta(s)\alpha(t)\delta(t)$ for $s, t \in \Gamma$. It follows that the group $\Delta = \langle \delta(\Gamma) \rangle \leq G_{x_0}$ is normalized by $\alpha(\Gamma) = \Gamma$. Hence the tree Y of fixed points of Δ contains $[x_0, \varepsilon)$ and is Γ-invariant. Since $\Gamma \cdot [x_0, \varepsilon) = X$, clearly we have $\Delta = \{1\}$, i.e., $\delta(\Gamma) = \{1\}$, whence:

(24)   For $\alpha \in \mathrm{Aut}^{\mathrm{filt}}(\Gamma)$ we have $\rho(\alpha) \in N_{x_0}$, and $\alpha = ad_\Gamma(\rho(\alpha))$.

On the other hand, for $g \in N_{x_0}, s \in \Gamma$, and $r \geq 0$, we have $\rho(ad_\Gamma(g))(sx_r) = gsg^{-1}x_r = gsx_r$, whence

(25) $\rho(ad_\Gamma(g)) = g$ for $g \in N_{x_0}$.

From (24) and (25) we deduce

(26)   inverse isomorphisms $N_{x_0} \overset{ad_\Gamma}{\underset{\rho}{\rightleftarrows}} \mathrm{Aut}^{\mathrm{filt}}(\Gamma)$, and so, from (21),

$$\bigcap_{r \geq 0} N_G(\Gamma_r) = Z_G(\Gamma) \times N_{x_0} \cong Z_G(\Gamma) \times \mathrm{Aut}^{\mathrm{filt}}(\Gamma).$$

In view of (19), we further have

(27) $$N_G(\Gamma) \cong \Gamma \rtimes_{\Gamma_0 = ad_\Gamma(\Gamma_0)} \mathrm{Aut}^{\mathrm{filt}}(\Gamma), \quad \text{and}$$
$$N_G(\Gamma)/\Gamma \cong \mathrm{Aut}^{\mathrm{filt}}(\Gamma)/ad_\Gamma(\Gamma_0).$$

We next describe $G = \mathrm{Aut}(X)$. Note that X and G are determined by

(28) $I(\Gamma \backslash\backslash X) = I(G \backslash\backslash X) = \underset{}{\circ} \frac{1 \quad q_1 \quad 1 \quad q_2 \quad 1 \quad q_3 \quad 1 \quad q_4 \quad 1}{\circ \quad \circ \quad \circ \quad \circ \quad \circ} \ldots.$

Recall (18): $G = \Gamma \cdot G_{x_0}$, $\Gamma \cap G_{x_0} = \Gamma_0$, and G_{x_0} fixes all x_r. We can schematically picture X as

(29) $X =$

where

(30) $$I(G_{x_0} \backslash\backslash Y_r) = \underset{y_0}{\circ} \frac{1 \quad q_1 \quad 1 \quad q_2 \quad 1}{\underset{y_1}{\circ} \quad \underset{y_2}{\circ}} \ldots \frac{q_{r-1} \quad 1 \quad q_r - 1}{\underset{y_{r-1}}{\circ} \quad \underset{y_r}{\circ}}$$

($y_r =$ image of the root x_r of Y_r).

Moreover

(31) $$G_{x_0} \cong \prod_{r \geq 1} \mathrm{Aut}(Y_r, x_r).$$

Here $\mathrm{Aut}(Y_r, x_r)$ denotes the automorphism group of the (finite) rooted tree (Y_r, x_r). It has the following structure (cf. [BORT], Ch. III)

$$(32) \qquad \mathrm{Aut}(Y_r, x_r) \cong S_{q_1}^{Q_1} \rtimes S_{q_2}^{Q_2} \rtimes \cdots S_{q_{r-1}}^{Q_{r-1}} \rtimes S_{q_r - 1},$$

an iterated wreath product, where

$$(33) \qquad Q_s = q_{s+1} \cdots q_{r-1} \cdot (q_r - 1) \quad (1 \le s < r).$$

and S_{q_s} denotes the symmetric group on q_s elements. More generally, writing

$$(34) \qquad X =$$

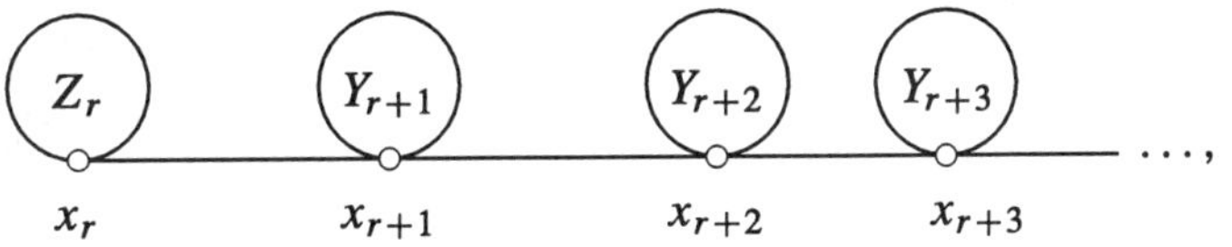

with

$$Z_r =$$

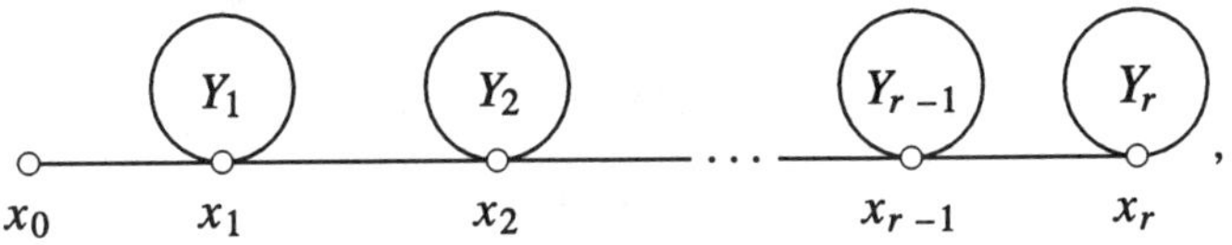

we have

$$(35) \qquad G_{x_r} = \mathrm{Aut}(Z_r, x_r) \times \prod_{s > r} \mathrm{Aut}(Y_s, x_s),$$

with

$$(36) \qquad I(G_{x_r} \backslash\backslash Z_r) = \underset{z_0}{\overset{1}{\circ}} \frac{q_1}{} \underset{z_1}{\overset{1}{\circ}} \frac{q_2}{} \underset{z_2}{\overset{1}{\circ}} \cdots \frac{q_{r-1}}{} \underset{z_{r-1}}{\overset{1}{\circ}} \frac{q_r}{} \underset{z_r}{\overset{}{\circ}}.$$

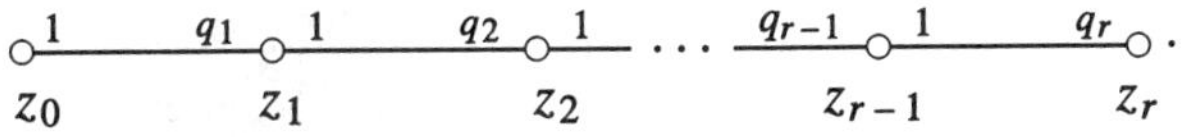

In this case $\mathrm{Aut}(Z_r, x_r)$ has the same structure as in (32), but with $q_r - 1$ in (32) replaced here by q_r. Moreover,

$$(37) \qquad G = \bigcup_{r \ge 0} G_{x_r} \quad \text{(ascending union)}.$$

9.10   Parabolic ray lattices

Consider a *parabolic ray*

$$(1) \qquad (A, i) =$$

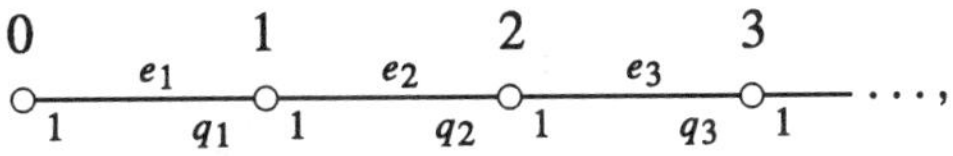

with indices $q_r = i(\bar{e}_r) \geq 2$ for almost all $r \geq 1$. Let

$$(2) \qquad\qquad X = \widehat{(A, i, 0)},$$

the (parabolic) universal covering tree; X has bounded degree iff the q_r are bounded. Let

$$(3) \qquad\qquad G = \mathrm{Aut}(X); \quad \text{clearly} \ (A, i) = I(G\backslash\backslash X).$$

Specifying $\Gamma \leq G$ with $\Gamma\backslash X = G\backslash X$ is equivalent to giving a chain of groups,

$$(4) \qquad\qquad \Gamma_0 < \Gamma_1 < \Gamma_2 < \Gamma_3 < \cdots ; \quad \Gamma = \bigcup_r \Gamma_r,$$

such that

(5)　　(i)　$[\Gamma_r : \Gamma_{r-1}] = q_r$ for $r \geq 1$, and
　　　　(ii)　Γ_0 contains no non-trivial normal subgroup of Γ.

In this case, Γ is an X-lattice iff

$$(6) \qquad\qquad \Gamma_0 \ \text{(and hence all} \ \Gamma_r) \ \text{are finite.}$$

In fact we shall often restrict attention to the case,

$$(6)_0 \qquad\qquad \Gamma_0 = \{1\},$$

which assures both (6) and (5)(ii). By (9.9)(16), we have an isomorphism

$$(7) \qquad\qquad \zeta : \left(\bigcap_{r \geq 0} N_\Gamma(\Gamma_r)\right) / \Gamma_0 \longrightarrow Z_G(\Gamma).$$

Choosing (4) with all Γ_r normal in Γ (hence $\Gamma_0 = \{1\}$), e.g., taking Γ abelian—we can make $Z_G(\Gamma) \cong \Gamma$ as large as possible. On the other hand, choosing (4) so that $N_\Gamma(\Gamma_0) = \Gamma_0$, we can make $Z_G(\Gamma) \cong \Gamma_0$, as small as possible.

To study $N_G(\Gamma)$ we use the isomorphisms of (9.9)(27),

$$(8) \qquad \begin{aligned} N_G(\Gamma) &\cong \Gamma \rtimes_{\Gamma_0 = ad_\Gamma(\Gamma_0)} \mathrm{Aut}^{\mathrm{filt}}(\Gamma), \quad \text{and} \\ N_G(\Gamma)/\Gamma &\cong \mathrm{Aut}^{\mathrm{filt}}(\Gamma)/ad_\Gamma(\Gamma_0). \end{aligned}$$

The following notation will be convenient. For groups $U \leq V$, write

$$\mathrm{Aut}(V; U) = \{\alpha \in \mathrm{Aut}(V) |\ \alpha U = U\}.$$

More generally, if $(U_r)_r$ is a family of subgroups $U_r \leq V$, put

$$\mathrm{Aut}(V; (U_r)_r) = \bigcap_r \mathrm{Aut}(V, U_r).$$

For example, in (8) above we have

$$\text{Aut}^{\text{filt}}(\Gamma) = \text{Aut}(\Gamma; (\Gamma_r)_{r \geq 0}).$$

There are many ways to make $\text{Aut}^{\text{filt}}(\Gamma)$ infinite, e.g., when Γ is abelian, such as a vector space over a finite field. We discuss now ways of making it finite. To this end consider the condition, for $r > 0$:

$(9)_r$
$\quad$ (i) $Z_{\Gamma_{r-1}}(\Gamma_r) = \{1\}, \quad$ and
$\quad$ (ii) $\text{Aut}(\Gamma_r; \Gamma_{r-1}) = ad_{\Gamma_r}(\Gamma_{r-1})$.

This is equivalent to the condition

$(10)_r \qquad ad_{\Gamma_r} : \Gamma_{r-1} \xrightarrow{\cong} \text{Aut}(\Gamma_r; \Gamma_{r-1})$ is an isomorphism.

Now assuming, for some $r_0 > 0$,

$(11)_{r_0} \qquad\qquad (9)_r$ holds for all $r > r_0$,

then it follows easily that we have an isomorphism

$(12) \qquad\qquad ad_\Gamma : \left(\bigcap_{r < r_0} N_{\Gamma_{r_0}}(\Gamma_r) \right) \xrightarrow{\cong} \text{Aut}^{\text{filt}}(\Gamma)$

In fact, if $\alpha \in \text{Aut}^{\text{filt}}(\Gamma)$ and $r > r_0$ then, by $(11)_{r_0}$ and $(10)_r$, $\alpha | \Gamma_r = ad_{\Gamma_r}(g_{r-1})$ for a unique $g_{r-1} \in \Gamma_{r-1}$. Since $ad(g_{r-1})$ and $ad(g_r)$ both agree with α in Γ_r, and $Z(\Gamma_r) = \{1\}$ by $(9)_{r+1}(i)$, we have $g_{r-1} = g_r \in \Gamma_{r-1}$. Thus, for all $r > r_0$, all g_{r-1} coincide with $g = g_{r_0} \in \Gamma_{r_0}$, and $\alpha = ad_\Gamma(g)$. Moreover g normalizes all $\Gamma_r (r \geq 0)$, since $\alpha \in \text{Aut}^{\text{filt}}(\Gamma)$. Note that (12) exhibits $N_G(\Gamma)/\Gamma$ as the finite group, $(\bigcap_{0 < r < r_0} N_{\Gamma_{r_0}}(\Gamma_r))/\Gamma_0$. The smallest possibility for this group occurs when r_0 is the least r for which $\Gamma_r \neq \{1\}$, in which case we have:

(13)$\quad$ If r_0 is the least r for which $\Gamma_r \neq \{1\}$, and if $(11)_{r_0}$ holds, then $N_G(\Gamma)/\Gamma \cong$
$\qquad \Gamma_{r_0}/\Gamma_0 = \Gamma_{r_0}$ if $r_0 > 0$ (so $\Gamma_0 = \{1\}$), and $= \{1\}$ if $r_0 = 0$.

Finally, we observe from (6.18) and (6.1)(14) that:

(14)$\quad$ If Γ is not residually finite then the commensurator $C_G(\Gamma)$ is not dense in G.
$\qquad$ If $\hat{\Gamma} = \{1\}$, then $C_G(\Gamma) = N_G(\Gamma)$.

In fact, (6.18) says here, more precisely, that if $\Gamma \leq H = G_H \leq G$, and if $\overline{C_H(\Gamma)} = H$, then Γ is residually finite.

9.11$\quad$ **Example.** The *factorial ray* is

(1)$\quad (A, i) = $

$$\underset{1}{\overset{0}{\circ}} \xrightarrow[1]{e_1} \underset{1}{\overset{1}{\circ}} \xrightarrow[2]{e_2} \underset{2}{\overset{2}{\circ}} \xrightarrow[3]{e_3} \underset{3}{\overset{3}{\circ}} \cdots \underset{n-1}{\overset{n-1}{\circ}} \xrightarrow[n]{e_n} \underset{n}{\overset{n}{\circ}} \cdots$$

$$i(e_n) = 1, \quad i(\bar{e}_n) = n \qquad \forall \; n \geq 1.$$

(In the notation of (9.9) and (9.10), $q_n = n$). For any (finite) grouping

(2)

$$\Gamma_0 = \{1\} \quad \Gamma_1 \quad \Gamma_2 \quad \Gamma_3 \quad \cdots \quad \Gamma_{n-1} \quad \Gamma_n \quad \cdots, \quad \Gamma = \bigcup_n \Gamma_n,$$

we have

(3)

$$\mathrm{Vol}(\Gamma \backslash\backslash X) = \sum_{n \geq 0} \frac{1}{|\Gamma_n|} = \sum_{n \geq 0} \frac{1}{n!} = e.$$

A *product grouping* is of the form

(4)

$$\Gamma_n = C_1 \times \cdots \times C_n,$$

$$C_n = \text{any group of order } n.$$

(Note that $\Gamma_0 = \{1\}$, the "empty product"!) From (9.9)(17)(c) we have

(5)

$$Z_G(\Gamma) \cong \Gamma, \text{ since } \Gamma_n \lhd \Gamma \; \forall n, \text{ and } \Gamma_0 = \{1\}.$$

By (9.10)(8).

$$N_G(\Gamma)/\Gamma \cong \mathrm{Aut}^{\mathrm{filt}}(\Gamma).$$

It is easily seen that

(6)

$$\mathrm{Aut}^{\mathrm{filt}}(\Gamma) = U \times \left(\prod_{n \geq 1} \mathrm{Aut}(C_n) \right),$$

$$U \cong \prod_{n \geq 1} Hom(C_n, Z(\Gamma_{n-1})).$$

Note that U will be trivial if either $Z(C_n) = \{1\} \; \forall n$, or $C_n^{ab} = \{1\} \; \forall n$. It follows in particular from (6) that

(7)

$$N_G(\Gamma)/\Gamma \; (\cong \mathrm{Aut}^{\mathrm{filt}}(\Gamma)) \text{ is infinite.}$$

A *divisible grouping* is obtained by taking

(8)

$$\Gamma_n = \frac{1}{n!}\mathbf{Z}/\mathbf{Z} < \Gamma = \mathbf{Q}/\mathbf{Z}$$

In this case again $\Gamma_0 = \{0\}$ so

(9)

$$Z_G(\Gamma) \cong \Gamma \quad \text{since } \Gamma_n \lhd \Gamma \; \forall n.$$

Moreover, since each Γ_n is a characteristic subgroup of Γ, we have

(10)
$$N_G(\Gamma)/\Gamma \cong \mathrm{Aut}^{\mathrm{filt}}(\Gamma) = \mathrm{Aut}(\Gamma)$$
$$\cong \hat{\mathbf{Z}}^\times,$$

the group of units in the profinite completion $\hat{\mathbf{Z}}$ of $\mathbf{Z}$. Note that Γ, being divisible, has trivial profinite completion, the farthest from being residually finite. Hence, by (9.10)(14),

(11) $C_G(\Gamma)$ is not dense in G. In fact, $C_G(\Gamma) = N_G(\Gamma)$ is a closed proper subgroup of G.

Next consider the *symmetric grouping* of (A, i),

$$\Gamma_n = S_n \text{ (symmetric group), and}$$
$$\Gamma'_n = A_n \text{ (alternating group) } (\Gamma_0 = \{1\} = \Gamma'_0).$$

We identify Γ_n with the subgroup of Γ_{n+1} fixing $n + 1$. Then

(12)
$$\Gamma := \bigcup_n \Gamma_n = S_\infty, \quad \text{and} \quad \Gamma' := \bigcup_n \Gamma'_n = A_\infty$$

are lattices on X, with $\Gamma \backslash X = A$, and

(13) Γ' is a simple group of index 2 in Γ.

We recall now some classical facts (cf. Miller, Blichtfeldt, and Dickson, *Theory and Applications of Finite Groups*, Chapter VII, Section 65, and J. Rotman, *An Introduction to the Theory of Groups*, 4th ed., Springer-Verlag, Chapter 7, pp. 156–162).

(14)   (S) $ad_{S_n} : S_n \longrightarrow \mathrm{Aut}(S_n)$ is injective for $n \neq 2$, and surjective for $n \neq 6$;

$$\mathrm{Aut}(S_6) = ad(S_6) \rtimes \langle \varepsilon \rangle, \quad \varepsilon^2 = Id.$$

(A) $A_n (= (S_n, S_n))$ is a characteristic subgroup of S_n. Restriction,

$$res : \mathrm{Aut}(S_n) \longrightarrow \mathrm{Aut}(A_n),$$

is surjective for all $n \geq 1$, and injective for $n \neq 3$;

$$Ker(\mathrm{Aut}(S_3) = ad_{S_3}(S_3) \xrightarrow{res} \mathrm{Aut}(A_3)) = ad_{S_3}(A_3).$$

(N) $N_{S_{n+1}}(S_n) = S_n$ for $n \geq 2$, and $N_{A_{n+1}}(A_n) = A_n$ for $n \geq 3$.

Property (N) follows because, in each case, the normalizer fixes the unique fixed point, $n + 1$, of the normalized subgroup.

We now calculate the centralizers, normalizers, and commensurators in G of Γ and Γ'. First recall the "horosphere sequence" ((9.9)(5)) of X:

(15)
$$X_0 \xrightarrow{q} X_1 \xrightarrow{q} X_2 \xrightarrow{q} X_3 \longrightarrow \cdots$$
$$X_n = \Gamma \cdot x_n, \quad \Gamma_{x_n} = \Gamma_n, \quad \text{so} \quad X_n \cong \Gamma/\Gamma_n, \quad \text{and}$$
$$q(sx_n) = sx_{n+1} \quad \text{for} \quad n \geq 0.$$

Let

$$\Gamma_2 = S_2 = \langle \sigma \rangle, \quad \sigma = (1,2); \quad \Gamma'_3 = A_3 = \langle \gamma \rangle, \quad \gamma = (1,2,3).$$

Note that, from (14)(N) and (9.9)(16), we have

(16)
$$\text{(a)} \qquad \bigcap_{n \geq 0} N_\Gamma(\Gamma_n) = S_2 = \Gamma_2 \cong Z_G(\Gamma);$$
$$\text{(a')} \qquad \bigcap_{n \geq 0} N_{\Gamma'}(\Gamma'_n) = A_3 = \Gamma'_3; \quad \text{and}$$
$$\text{(b)} \qquad \Gamma_3 = S_3 \text{ normalizes } A_n \text{ for all } n \geq 0.$$

We define a new action of S_3 on X as follows:

(17) For $g \in S_3$, $s \in \Gamma'$, $i = 0, 1$, and $n \geq 0$, define $\bar{g}(s\sigma^i x_n) = sg^{-1}\sigma^i x_n$.

First note that $\bar{g}$ is well defined. For $n = 0$, this follows since then Γ acts freely on X_0. For $n > 0$, this follows from (16)(b) and the fact that $\sigma^i x_n = x_n$. Clearly $g \mapsto \bar{g}$ is a homomorphism. Moreover $q\bar{g}(s\sigma^i x_n) = q(sg^{-1}\sigma^i x_n) = sg^{-1}\sigma^i x_{n+1} = \bar{g}(s\sigma^i x_{n+1})$, so $\bar{g} \in G = \text{Aut}(X)$. Thus:

(18)
(a) The map $g \to \bar{g}$ of (17) defines an isomorphism $S_3 \xrightarrow{\cong} \bar{S}_3 < G$.
(b) $\bar{S}_3$ acts freely on X_0, and fixes x_n for $n \geq 3$.
(c) $\bar{S}_3$ centralizes Γ'.
(d) $\bar{S}_2$ centralizes Γ.

Assertions (a), (b), and (c) are clear from (17) and the discussion above. For $t = s\sigma^i \in \Gamma$ as in (17), and $n \geq 0$, $\bar{\sigma}(tx_n) = s\sigma\sigma^i x_n = s\sigma^i \sigma x_n = t\bar{\sigma}x_n$, whence (d) follows immediately.

Now from (16)(a) and (9.9)(16) we have

(19)
$$Z_G(\Gamma) = \bar{S}_2.$$

Further, from (9.9)(26) and (27),

$$N_G(\Gamma) = \Gamma \rtimes N_G(\Gamma)_{x_0}, \quad \text{and}$$
$$ad_\Gamma : N_G(\Gamma)_{x_0} \xrightarrow{\cong} \text{Aut}^{\text{filt}}(\Gamma)$$

From $(14)(S)$ and (N) we have conditions $(9)_n$ and $(10)_n$ of (9.10) for $n > 6$, whence by $(9.10)(12)$, an isomorphism of $\mathrm{Aut}^{\mathrm{filt}}(\Gamma)$, with $\bigcap_{0<n<6} N_{S_6}(S_n) = S_2$, by $(14)(N)$ again. Thus

(20)
$$N_G(\Gamma) = \Gamma \rtimes N_G(\Gamma)_{x_0}, \quad \text{and}$$
$$N_G(\Gamma)_{x_0} = \langle \sigma^* \rangle \cong S_2, \quad \text{with}$$
$$\sigma^*(tx_n) = \sigma t \sigma x_n \quad \text{for } t \in \Gamma, \ n \geq 0.$$

Now for Γ'. From (13) we see that Γ' must be normalized by the commensurator $C_G(\Gamma)$. Hence,

(21)
$$N := N_G(\Gamma') = C_G(\Gamma) = C_G(\Gamma').$$

A fundamental domain for Γ' on X is

(22)

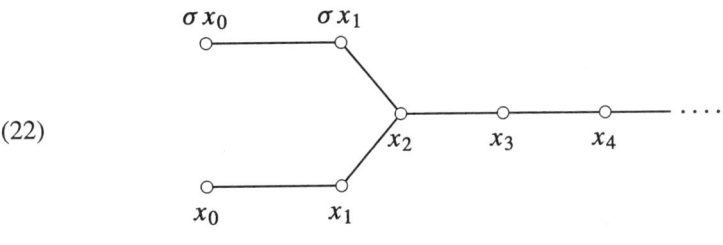

with $\Gamma'_{x_n} = \Gamma'_{\sigma x_n} = \Gamma'_n = A_n$ for all $n \geq 0$.

It follows that $Z_G(\Gamma')$ consists of all Γ'-automorphisms of the diagram of Γ'-sets

(23)
$$
\begin{array}{ccc}
\Gamma'\sigma x_0 & \xrightarrow{\ q\ } & \Gamma'\sigma x_1 \\
 & & \searrow{\scriptstyle q} \\
 & & \qquad \Gamma' x_2 \to \Gamma' x_3 \to \cdots \\
 & & \nearrow{\scriptstyle q} \\
\Gamma' x_0 & \xrightarrow[\ q\ ]{} & \Gamma'\sigma x_1
\end{array}
$$

A straightforward (and moderately laborious) calculation shows that

(24)
$$Z_G(\Gamma') = \bar{S}_3 \cong S_3.$$

Now for $N = N_G(\Gamma')$, since Γ on X_0 is simply transitive we have

(25)
$$N = \Gamma \cdot N_{x_0}, \quad \Gamma \cap N_{x_0} = \{1\}.$$

Since, by $(18)(b)$ and (24), $N_{x_0} \cap Z_G(\Gamma') = \{1\}$, and N_{x_0} fixes x_n for all $n \geq 0$, we have a monomorphism

(26)
$$ad_{\Gamma'} : N_{x_0} \longrightarrow \mathrm{Aut}^{\mathrm{filt}}(\Gamma').$$

From the conditions of (14) we deduce an isomorphism

$$(27) \qquad ad_{\Gamma'} : S_3 = \bigcap_{n \geq 0} N_\Gamma(\Gamma'_n) \overset{\cong}{\longrightarrow} \mathrm{Aut}^{\mathrm{filt}}(\Gamma').$$

Thus we obtain an isomorphism

$$
\begin{aligned}
(28) \qquad & S_3 \longrightarrow S_3^* = N_{x_0}, \quad g \mapsto g^*, \\
& g^*(s\sigma^i x_n) = gsg^{-1}\sigma^i x_n, \quad \text{for } s \in \Gamma', \; i = 0, 1, \; n \geq 0. \\
& \text{In the notation of (17), } g^* = g\bar{g} \text{ for } g \in S_3 = \Gamma_3 < G.
\end{aligned}
$$

Hence, since $\bar{S}_3 = Z_G(\Gamma') \triangleleft N$,

$$(29) \qquad N = \Gamma \ltimes \bar{S}_3 .$$

Note that for $g = \sigma$, the notation σ^* in (28) agrees with that in (20). The action of Γ on $\bar{S}_3$ factors through Γ / Γ', and σ acts by $ad_{\bar{S}_3}(\bar{\sigma})$. Putting

$$
\begin{aligned}
(30) \qquad & \Gamma^{(*)} := \Gamma' \cdot \langle \sigma^* \rangle \cong \Gamma, \quad \text{we have} \\
& N = \Gamma^{(*)} \times \bar{S}_3.
\end{aligned}
$$

Finally, since N/Γ is finite, we have $C_G(N) = C_G(\Gamma) = N$, by (21). Therefore:

$$
\begin{aligned}
(31) \qquad & N = N_G(\Gamma') = C_G(\Gamma') = C_G(\Gamma) = C_G(N) \text{ is an } X\text{-lattice}, \\
& N \backslash X = \Gamma \backslash X = G \backslash X, \quad Z_G(N) = \{1\}, \quad \text{and} \\
& \mathrm{Vol}(N \backslash\backslash X) = \frac{1}{6}\mathrm{Vol}(\Gamma \backslash\backslash X) = \frac{e}{6}.
\end{aligned}
$$

Moreover N has profinite completion $\hat{N} \cong S_3$.

9.12 **Example.** The *q-power ray* is

$$(1) \qquad i(e_n) = 1, \quad i(\bar{e}_n) = q \quad \text{for all } n \geq 1.$$

Here q is some integer ≥ 2. For any finite faithful grouping we have

(3) $$\mathrm{Vol}(\Gamma\backslash\backslash X) = \sum_{n\geq 0}\frac{1}{|\Gamma_n|} = \frac{1}{|\Gamma_0|}\sum_{n\geq 0}q^{-n} = \frac{q}{|\Gamma_0|(q-1)}.$$

Note that $X(= (\widehat{A,i,0}))$ is a parabolic tree of *bounded degree*. In fact all vertices of X have degree $q + 1$, except for the "leaves" (= terminal vertices, lying over the vertex 0 of A), which have degree 1.

For certain values of q we can choose (2) to be the *PNeumann grouping* (cf. Appendix [PN]). This is formed from a simple inn-exact group S of order m, and a transitive rigid S-set V of size $r > 1$, and we then have (Appendix [PN], (2.1)(2)),

(4) $$q = m^{r^2+1} \quad \text{and} \quad |\Gamma_0| = m^2, \quad \text{so that}$$

(5) $$\mathrm{Vol}(\Gamma\backslash\backslash X) = \frac{m^{r^2+1}}{m^2(m^{r^2+1}-1)} = \frac{m^{r^2-1}}{m^{r^2+1}-1}.$$

We have

(6) $$Z_G(\Gamma) = \{1\},$$

because, by (9.10)(7), $Z_G(\Gamma) \cong (\bigcap_{n\geq 0} N_\Gamma(\Gamma_n))/\Gamma_0 = \{1\}$, by Appendix [PN], (2.1)(6)(a). Further,

(7) $$N_G(\Gamma) = \Gamma,$$

because, by (9.10)(8), $N_G(\Gamma)/\Gamma \cong \mathrm{Aut}^{\mathrm{filt}}(\Gamma)/ad_\Gamma(\Gamma_0) = \{1\}$, by Appendix [PN], (2.1)(6)(b).

Thus, on a parabolic tree of bounded degree, we have a lattice Γ satisfying (6) and (7).

9.13   Parabolic lattices with all horospheres infinite

Let X be a parabolic tree with (unique) end ε_X, with a lattice Γ. Then ((9.7))

(1) $$(A, \varepsilon) := \Gamma\backslash(X, \varepsilon_X)$$

is again a parabolic tree; put $(A, i) = I(\Gamma\backslash\backslash X)$, and let

(3) $$E^+A = \text{the set of edges of } A \text{ directed toward } \varepsilon.$$

For $e \in E^+A$ we have

(4) $$i(e) = 1, \quad \text{so } \Delta(e) := \frac{i(\bar{e})}{i(e)} = i(\bar{e}) = \Delta(\bar{e})^{-1}.$$

Pick a base point $a_0 \in VA$. For any $a \in VA$ we have (cf. (2.6))

(5) $\frac{\Delta a}{\Delta a_0} = \Delta(e_1) \cdots \Delta(e_n)$ $(= (\frac{\Delta a_0}{\Delta a})^{-1})$, where $(e_1, \ldots, e_n)$ is any edge path from a_0 to a.

Since Γ is an X-lattice,

(6) $$\text{Vol}_{a_0}(A, i) := \sum_{a \in VA} \frac{\Delta a_0}{\Delta a} < \infty.$$

and (A, i) has "bounded denominators" (cf. (2.6)(18)):

(7)   The numbers $\frac{\Delta a}{\Delta a_0} \in \mathbf{Q}$ have a common denominator.

In view of (4), $\frac{\Delta a}{\Delta a_0}$ increases (multiplicatively) along paths toward ε, and so similarly decreases along paths away from ε. The latter are all finite, as they must be to allow for the convergence of (6). But we can still ask whether A admits paths that go arbitrarily "far" back, away from ε? More precisely, consider the horosphere sequence of A,

(8) $\cdots \longrightarrow A_{-1} \xrightarrow{p} A_0 \xrightarrow{p} A_1 \xrightarrow{p} A_2 \longrightarrow \cdots$,

indexed so that $a_0 \in A_0$. Then we are asking whether we can make $A_{-n} \neq \emptyset$ for arbitrarily large n. Note that, were this to happen, then $A_n \neq \emptyset$ $\forall n \in \mathbf{Z}$, and even more: Since all paths away from ε are finite, this could happen only if

(9) A_n is infinite $\forall n \in \mathbf{Z}$.

This seems unlikely to happen in light of conditions (6) and (7). We shall nonetheless construct an example satisfying (9). In view of (2.6)(19), it suffices to show that:

(10)   There exists a parabolic edge-indexed tree (A, i, ε) satisfying (6), (7), and (9).

For the construction we start with a countable set Y and a map

        $t : Y \longrightarrow Y$ satisfying, $\forall y, y' \in Y$,

(11)   (a) $0 < |t^{-1}(y)| < \infty$ (t is surjective, with finite fibers),

            (b) $t^n(y) = t^n(y')$ for $n \gg 0$ $\left(\varinjlim(Y, t) = \{\varepsilon\}\right)$.

For example, fix an integer $m > 1$, let Y be all (complex) roots of unit y of order dividing some power m^n of m, and define $t(y) = y^m$. Then (a) $|t^{-1}(y)| = m$, and (b) $t^n(y) = 1$ for $n \gg 0$, for all $y \in Y$.

Now the parabolic tree A that we seek will be defined by its horosphere sequence (to be constructed)

(12)
                $\cdots \longrightarrow A_{-1} \xrightarrow{t} A_0 \xrightarrow{t} A_1 \xrightarrow{t} A_2 \longrightarrow \cdots$,   with
        $\varinjlim A_n = \{\varepsilon\}$,   a point,   and $\varprojlim A_n = \emptyset$.

In fact we then have $VA = \coprod_{n \in \mathbf{Z}} A_n$ and $E^+A = \{e_a = (a, t(a)) | a \in VA\}$, with $\partial_0 e_a = a$, $\partial_1 e_a = t(a)$. We start with $n \geq 0$, by taking $A_{\geq 0}$ to be the tree defined by the sequence

(12) $\qquad (A_0 \xrightarrow{t} A_1 \xrightarrow{t} A_2 \xrightarrow{t} \cdots) = (Y \xrightarrow{t} Y \xrightarrow{t} Y \xrightarrow{t} \cdots).$

Next, we define functions, for $n \geq 0$,

$$v_n : A_n \longrightarrow \mathbf{N} \text{ such that :}$$

(13)
$$\begin{aligned} &\text{(a)} \quad v_0(a_0) = 0, \text{ where } a_0 \in A_0 \text{ is a chosen base point;} \\ &\text{(b)} \quad v_n \text{ is injective; and} \\ &\text{(c)} \quad v_n(a) < v_{n+1}(t(a)) \quad \forall a \in A_n. \end{aligned}$$

Since $A_0 = Y$ is countable, we can list Y as a sequence, $y_0, y_1, y_2, \ldots$, and so define $v_0(y_n) = n$, taking $a_0 = y_0 \in A_0$. Inductively, suppose that $n > 0$ and v_i $(0 \leq i < n)$ are defined satisfying the relevant parts of (13). We define $v_n(y_m)$ inductively on m, so that $v_n(y_m) > v_n(y_i)$ for $i < m$, and $v_n(y_m) > v_{n-1}(y)$ for all $y \in t^{-1}(y_m)$, a finite set of inequalities by (11)(a).

The functions $v_n : A_n \longrightarrow \mathbf{N}(n \geq 0)$ assemble to define $v : VA_{\geq 0} \to \mathbf{N}$. We fix an integer $q > 1$ and set

(14) $\qquad\qquad\qquad \Delta(a) = q^{v(a)} \quad \forall a \in VA_{\geq 0}.$

We further index the edge

(15)
$$\begin{aligned} &e_a = (a, t(a)) \in E^+A_{\geq 0}, \\ &i(e_a) = 1, \quad \text{and} \quad i(\bar{e}_a) = \frac{\Delta(t(a))}{\Delta(a)} = q^{v(t(a)) - v(a)}. \end{aligned}$$

By (13)(c), $i(\bar{e}_a)$ is an integer > 1. Further

(16) For any edge path $\gamma = (e_1, \ldots, e_n)$ from a to b in $A_{\geq 0}$,

$$\Delta(\gamma) := \Delta(e_1) \cdots \Delta(e_n) = \frac{\Delta(b)}{\Delta(a)} = q^{v(b) - v(a)}.$$

Since $\Delta(a_0) = q^{v(a_0)} = 1$, by (13)(a), we have

(17) $\qquad\qquad \dfrac{\Delta(a)}{\Delta(a_0)} = \Delta(a) \qquad \forall a \in VA_{\geq 0}, \text{ so}$

(18)
$$\begin{aligned} \mathrm{Vol}_{a_0}(A_{\geq 0}, i) &= \sum_{a \in VA_{\geq 0}} \frac{1}{\Delta(a)} = \sum_{n \geq 0} V_n, \quad \text{where} \\ V_n &= \sum_{a \in A_n} q^{-v_n(a)}. \end{aligned}$$

Condition (13)(c), and the surjectivity of $t : A_{n-1} \longrightarrow A_n$ for all $n > 0$, implies that

$$(19) \qquad\qquad v_n(a) \geq n \qquad \forall a \in A_n.$$

From (19) and (13)(b) we conclude that

$$(20) \qquad V_n \leq \sum_{r \geq 0} q^{-n-r} = q^{-n}\left(\frac{1}{1 - 1/q}\right) = \frac{1}{(q-1)q^{n-1}}.$$

Feeding (20) into (18) we have

$$\text{Vol}_{a_0}(A_{\geq 0}, i) \leq \sum_{n \geq 0} \frac{1}{(q-1)q^{n-1}}$$

$$(21)$$

$$= \frac{q}{(q-1)}\left(\frac{1}{1 - 1/q}\right) = \left(\frac{q}{q-1}\right)^2.$$

Now we complete the construction of A as follows. Choose a sequence

$$(22) \qquad (a_{r,0})_{r \geq 0} \text{ in } A_0, \text{ so that } v_0(a_{r,0}) \geq 2r \text{ for } r > 0.$$

This is possible because v_0 is injective ((13)(b)). For each $r > 0$ define a linear tree

$$(23) \qquad B_r = \overset{a_{r,-r}}{\circ}\!\!-\!\!\overset{a_{r,1-r}}{\circ}\!\!-\cdots-\overset{a_{r,-2}}{\circ}\!\!-\!\!\overset{a_{r,-1}}{\circ}\!\!-\!\!\overset{a_{r,0}}{\circ}$$

of length r, which we attach to $A_{\geq 0}$ by indentifying $a_{r,0} \in VB_r$ with $a_{r,0} \in A_0$ in (22). With all of the B_r so attached we obtain the parabolic tree A, which may be schematically pictured as follows:

Thus, for $n > 0$,

$$(25) \qquad A_{-n} = \{a_{r,-n} \mid r \geq n\}, \quad \text{and} \quad t(a_{r,-n}) = a_{r,-(n-1)}.$$

We index the edge

(26) $e_{r,-n} = (a_{r,-n}, t(a_{r,-n})) :$ $i(e_{r,-n}) = 1$ and $i(\bar{e}_{r,-n}) = q.$

Thus,

$$(B_r, i) = \underset{a_{r,-r}}{\circ}\!\overset{1}{\rule{1.2cm}{0.4pt}}\!\overset{q}{\circ}\!\overset{1}{\rule{1.2cm}{0.4pt}}\!\circ\ \cdots\ \overset{q}{\rule{1.2cm}{0.4pt}}\!\overset{1}{\circ}\!\overset{q}{\rule{1.2cm}{0.4pt}}\!\underset{a_{r,0}}{\circ},$$

Hence, for $0 \le n \le r$,

(27)
$$\Delta(a_{r,-n}) = \frac{\Delta(a_{r,-n})}{\Delta(a_{r,0})} \cdot \Delta(a_{r,0}) = q^{-n}\Delta(a_{r,0}) = q^{d_r-n},$$

where $d_r = v_0(a_{r,0}).$

By (22), $d_r \ge 2r > r$, so $d_r - n \ge r > 0$ above, and so $\Delta(a_{r,-n})$ is an integer for all $0 \le n \le r$, whence the bounded denominators condition (7). For the finite volume condition (6) we have

$$\text{Vol}_{a_0}(A, i) = \text{Vol}_{a_0}(A_{\ge 0}, i) + V_-,$$

where the first term is finite by (21), and

(28)
$$V_- = \sum_{r>0}\sum_{n=1}^{r}\frac{1}{q^{d_r-n}} = \sum_{r>0}q^{-d_r}(q + q^2 + \cdots + q^r)$$
$$= \sum_{r>0}q^{1-d_r}\left(\frac{q^r - 1}{q - 1}\right) = \frac{q}{q-1}\sum_{r>0}\frac{q^r - 1}{q^{d_r}} \le \frac{q}{q-1}\left(\sum_{r>0}\frac{1}{q^{d_r-r}}\right)$$

By (22), $d_r \ge 2r$ so $d_r - r \ge r$, and so $V_- \le \frac{q}{q-1}(\sum_{r>0}q^{-r}) < \infty$, whence $\text{Vol}_{a_0}(A, i) < \infty$, as desired. One unpleasant feature of (A, i) and X is that X has unbounded degree (cf. (15)). We next give a variant on the construction that produces an example of bounded degree.

9.14   A bounded degree example

Set $Y = \{1, 2, 3, 4, 5, 6, 7, \ldots\}$ and

(1) $t : Y \longrightarrow Y,$ $t(2n - 1) = t(2n) = n\ \forall n.$

For $n \ge 1$ write $n = 2^s + i$, $s \ge 0$, $0 \le i < 2^s$, and define

$$v_n : Y \longrightarrow \mathbf{Z},$$

(2)
$$v_{2^s+i}(r) = \begin{cases} 2^s r + i, & r \text{ odd}; \\ 2^s r, & r \text{ even}. \end{cases}$$

$$(s \geq 0, \ 0 \leq i < 2^s)$$

We now make our horosphere sequence as in (9.13)(8). For $n \geq 1$ put $A_n = Y$, and define

(3)
$$p = p_n : A_n \longrightarrow A_{n+1}, \quad p_n = \begin{cases} t & \text{if } n + 1 = \text{a power of 2}, \\ Id_Y & \text{otherwise}. \end{cases}$$

(A_1, v_1) (A_2, v_2) (A_3, v_3) (A_4, v_4) (A_5, v_5) (A_6, v_6) (A_7, v_7) (A_8, v_8) (A_9, v_9) (A_{10}, v_{10})

1 ∘
     ↘   2 ∘ → 3 ∘
2 ∘ ↗   ↘
                  4 ∘ → 5 ∘ → 6 ∘ → 7 ∘
3 ∘ ↘ 4 ∘ → 4 ∘ ↗
4 ∘ ↗
                                ↘   8 ∘ → 9 ∘ → 10 ∘ ⋯
5 ∘ ↘
      6 ∘ → 7 ∘   ↗
6 ∘ ↗   ↘
             8 ∘ → 8 ∘ → 8 ∘ → 8 ∘
7 ∘ ↘ 8 ∘ → 8 ∘ ↗
8 ∘ ↗
9 ∘ ↘
      10 ∘ → 11 ∘
10 ∘ ↗   ↘
             12 ∘ → 13 ∘ → 14 ∘ → 15 ∘
11 ∘ ↘ 12 ∘ → 12 ∘ ↗
12 ∘ ↗   ↘
                               16 ∘ → 16 ∘ → 16 ∘ ⋯
13 ∘ ↘   ↗
      14 ∘ → 15 ∘
14 ∘ ↗   ↘
             16 ∘ → 16 ∘ → 16 ∘ → 16 ∘

We claim that

(4) For all $n \geq 1$ and $r \geq 1$, $v_{n+1}(p_n(r)) - v_n(r) = \begin{cases} 1, & (o), \\ 0, & (e). \end{cases}$

Here (o) and (e) denote the cases "r odd" and "r even," respectively. To see this, write $n = 2^s + i$, $s \geq 0$, $0 \leq i < 2^s$. Then

(2) $$v_n(r) = \begin{cases} 2^s r + i, & (o), \\ 2^s r, & (e). \end{cases}$$

Case $i < 2^s - 1$. Then $n + 1 = 2^s + (i + 1)$, $p_n = Id$, and

$$v_{n+1}(p_n(r)) = \begin{cases} 2^s r + i + 1, & (o), \\ 2^s r, & (e), \end{cases}$$

whence (4).

Case $i = 2^s - 1$. Then $n + 1 = 2^{s+1}$, $p_n = t$, by (3), and

$$v_{n+1}(p_n(r)) = 2^{s+1} t(r) = \begin{cases} 2^{s+1}(\frac{r+1}{2}) = 2^s r + i + 1, & (o), \\ 2^{s+1}(\frac{r}{2}) = 2^s r, & (e), \end{cases}$$

whence (4) again.

Now to index the parabolic tree $A_{\geq 1}$ defined by the horosphere sequence

$$A_1 \xrightarrow{P} A_2 \xrightarrow{P} A_3 \xrightarrow{P} A_4 \xrightarrow{P} \cdots$$

we index the edge $e_n(r) = (r, p_n(r))$ $\qquad (r \in A_n = Y)$ by

(5) $$i(e_n(r)) = 1$$
$$i(\bar{e}_n(r)) = q^{v_{n+1}(p(r)) - v_n(r)} = \begin{cases} q, & r \text{ odd}, \\ 1, & r \text{ even} \end{cases}$$

by (4). Thus the local picture at a vertex of $(A_{\geq 1}, i)$ looks like one of the following cases, with the end ε to the right:

degree 2,

degree $q + 1$,

degree $q + 2$.

Hence

(6)   All vertices of $(A_{\geq 1}, i)$ have degree $\leq q + 2$.

Define $v : VA \to \mathbf{Z}$ by $v|A_n = v_n$, and

(7)
$$\Delta : VA \to \mathbf{Z},$$
$$\Delta(a) = q^{v(a)}.$$

From (5) above we see that

(8)   For $e \in EA_{\geq 1}$, $\dfrac{\Delta(\partial_1 e)}{\Delta(\partial_0 e)} = \dfrac{i(\bar{e})}{i(e)} = q^{v(\partial_1 e) - v(\partial_0 e)}$.

Moreover, letting $a_0 =$ the vertex $1 \in A_1$, we have $\Delta(a_0) = q^{v_1(1)} = q$. It follows from this and (8) that

(9) $\dfrac{\Delta(a)}{\Delta(a_0)} = q^{v(a)-1} \qquad \forall a \in VA.$

Hence

$$\mathrm{Vol}_{a_0}(A_{\geq 1}, i) = q \sum_{n \geq 1} \sum_{r \geq 1} q^{-v_n(r)}$$

$$= q \left(\sum_{s \geq 0} \sum_{0 \leq i < 2^s} \sum_{m \geq 1} \left(\frac{1}{q^{2^s(2m-1)+i}} + \frac{1}{q^{2^s \cdot 2m}} \right) \right).$$

Now $\sum_{m \geq 1} \frac{1}{q^{2^s \cdot 2m}} = \frac{1}{q^{2^{s+1}} - 1}$, and

$$\sum_{m \geq 1} \frac{1}{q^{2^s(2m-1)+i}} = q^{2^s - i} \sum_{m \geq 1} \frac{1}{q^{2^s \cdot 2m}} = \frac{q^{2^s - i}}{q^{2^{s+1}} - 1}.$$

Thus

$$\mathrm{Vol}_{a_0}(A_{\geq 1}, i) = q \left[\sum_{s \geq 0} \sum_{0 \leq i < 2^s} \left(\frac{q^{2^s - i}}{q^{2^{s+1}} - 1} + \frac{1}{q^{2^{s+1}} - 1} \right) \right]$$

$$= q \sum_{s \geq 0} \frac{1}{q^{2^{s+1}} - 1} \left(\sum_{0 \leq i < 2^s} (q^{2^s - i} + 1) \right)$$

$$= q \sum_{s \geq 0} \frac{1}{q^{2^{s+1}} - 1} \left(\frac{q(q^{2^s} - 1)}{q - 1} + 2^s \right)$$

$$= \left(\frac{q^2}{q - 1} \right) \left(\sum_{s \geq 0} \frac{q^{2^s} - 1}{q^{2^{s+1}} - 1} \right) + q \left(\sum_{s \geq 0} \frac{2^s}{q^{2^{s+1}} - 1} \right)$$

$$\leq \left(\frac{q^2}{q-1} + q \right) \left(\sum_{s \geq 0} \frac{q^{2^s} - 1}{q^{2^{s+1}} - 1} \right) \quad (2^s \leq q^{2^s} - 1)$$

$$= \left(\frac{2q^2 - q}{q-1} \right) \left(\sum_{s \geq 0} \frac{1}{q^{2^s} + 1} \right)$$

$$\leq \left(\frac{2q^2 - q}{q-1} \right) \left(\sum_{m \geq 0} \frac{1}{q^m} \right) < \infty.$$

Now as in the previous section, we enlarge $A_{\geq 1}$ to A by attaching suitable linear trees to various vertices of A_1, as follows. We also indicate the value of Δ on the new vertices

$$A_{-4} \quad A_{-3} \quad A_{-2} \quad A_{-1} \quad A_0 \quad A_1 \quad A_2 \quad A_3 \quad A_4 \quad A_5 \quad A_6 \quad A_7 \quad A_8 \quad A_9$$

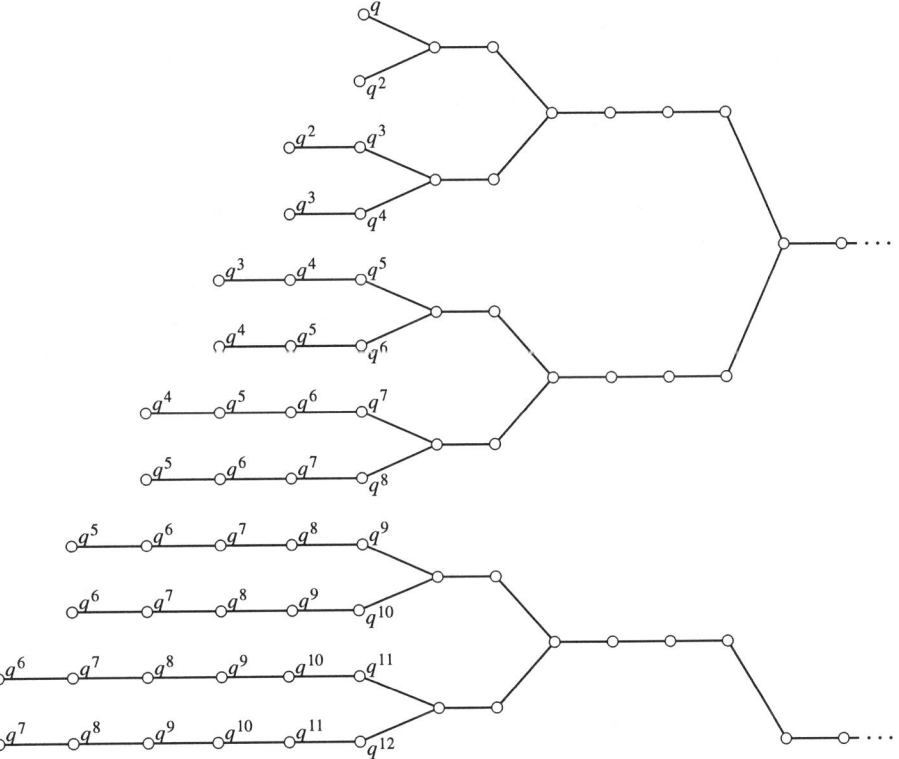

Clearly,

$$\mathrm{Vol}_{a_0}(A, i) = \mathrm{Vol}_{a_0}(A_{\geq 1}, i) + \sum_{n \geq 0} {}''\mathrm{Vol}(A_{-n})''$$

$$= \mathrm{Vol}_{a_0}(A_{\geq 1}, i) + \sum_{n \geq 0} \sum_{m \geq n+2} q^{-m}$$

$$= \mathrm{Vol}_{a_0}(A_{\geq 1}, i) + \sum_{n \geq 0} \frac{1}{q^{n+1}} \sum_{m \geq 1} \frac{1}{q^m}$$

$$= \mathrm{Vol}_{a_0}(A_{\geq 1}, i) + \sum_{n \geq 1} \frac{1}{q^n} \left(\frac{1}{q-1} \right)$$

$$= \mathrm{Vol}_{a_0}(A_{\geq 1}, i) + \frac{1}{(q-1)^2}.$$

9.15   Tree lattices that are simple groups must be parabolic

If Γ is a countable, locally finite, infinite simple group, then we can write Γ as the union of chain

$$\{1\} = \Gamma_0 < \Gamma_1 < \Gamma_2 < \Gamma_3 < \cdots$$

of finite subgroups. Then, as in (9.9), we can realize Γ as a parabolic tree lattice. Examples include the infinite alternating group A_∞ (cf. (9.11)) as well as simple algebraic groups over infinite algebraic extensions of finite fields.

We make here the observation that:

(1)   *If a tree lattice Γ is an infinite simple group, then Γ must be parabolic.*

Suppose, on the contrary, that an infinite simple group Γ is an X-lattice for some locally finite tree X, and that Γ is not parabolic. Then, by (9.7), $\ell(\Gamma) \neq 0$. Replacing X by the minimal subtree X_Γ (see (5.7)), we can assume that Γ acts minimally on X. Consider the quotient graph of groups

$$\mathbf{A} = (A, \mathcal{A}) = \Gamma \backslash\backslash X.$$

Since the free group $\pi_1(A)$ is a quotient of Γ (cf. (2.4)) we must have $\pi_1(A) = \{1\}$, i.e., A is a tree. Then every edge $e \in EA$ is separating, say

$$\mathbf{A} = \left(\mathbf{A}_0 \underset{a_0}{\bigcirc} \xrightarrow[e]{\mathcal{A}_e} \underset{a_1}{\bigcirc} \mathbf{A}_1 \right),$$

and we have

$$\Gamma = \Gamma_0 *_{\Gamma_e} \Gamma_1, \quad \text{where}$$

$$\Gamma_e = \mathcal{A}_e \quad \text{and} \quad \Gamma_i = \pi_1(\mathbf{A}_i, a_i) \ (i = 0, 1).$$

Since $\ell(\Gamma) \neq 0$, this amalgam must be non-trivial, by minimality. Moreover Γ_e is finite and Γ is infinite. If Γ_0 and Γ_1 are both finite, then Γ is virtually free (cf. (3.12) and [BK], (2.8)), contradicting simplicity. Thus

Γ_e is finite, $\Gamma_e \neq \Gamma_i$ $(i = 0, 1)$, and at least one of Γ_0 and Γ_1 is infinite.

Thus the impossibility of Γ being simple is a consequence of the following result, communicated to us by Ilya Rips.

(2)   *Consider an amalgam*

$$\Gamma = A *_C B$$

such that $C \neq B$ and $|C\backslash A/C| \geq 3$. Then Γ is not simple. (In fact, Γ is SQ-universal, i.e., every countable group is embeddable in a proper quotient of Γ.)

Note that the hypotheses are optimal since, for example, the simple group $\Gamma = PSL_2(\mathbf{Q}_p)$ is an amalgam $A *_C B$ with $A \cong B \cong PSL_2(\mathbf{Z}_p)$, C an Iwahori subgroup (triangular mod p), and $|C\backslash A/C| = |C\backslash B/C| = 2$.

Rips' explanation of (2) uses Small Cancellation Theory, as in Schupp (Small Cancellation Theory over Free Products with Amalgamation, *Math. Ann.*, **193** (1971), 255–264). Explicitly, let CaC and $Ca'C$ be distinct non-trivial double cosets in $C\backslash A/C$, and $b \in B - C$. Consider words in Γ of the form

$$w = a^{n_1} b a'^{n'_1} b a^{n_2} b a'^{n'_2} b a^{n_3} b a'^{n'_3} b \cdots .$$

When the exponents n_i, n'_i are suitably large one can apply Small Cancellation Theory to conclude that adding the relation $w = 1$ does not kill Γ, whence Γ is not simple.

9.16   Lattices on a product of two trees

It is interesting to contrast (9.15) with the situation for lattices on a product of two trees. Burger and Mozes [BM2, 4] have constructed remarkable lattices on a product of regular trees that are finitely presented torsion free simple groups of cohomological dimension 2. In fact, they are amalgams of two free groups.

This is part of a developing study of lattices on products of trees ([BG], [BM2, 4], [BMZ], [G1, 2], [M3]).

Chapter 10

Lattices of Nagao Type

10.1 Nagao Rays

A *Nagao ray of index* $(q_n)_{n \geq 0}$ is an indexed graph of the form

$$(A, i) =$$

(1)

$$i(e_0) = q_0 + 1, \quad i(\bar{e}_0) = q_1, \quad \text{and, for } n > 0,$$
$$i(e_n) = 1, \quad i(\bar{e}_n) = q_{n+1}.$$

We assume that

(2) $$q_0 \geq 1 \text{ and } q_n \geq 2 \text{ for almost all } n \geq 1.$$

A grouping $\mathbf{A}$ of (A, i) has the form

(3)

$$\Gamma_0^+ > \Gamma_0 < \Gamma_1 < \Gamma_2 < \Gamma_3 < \cdots,$$
$$[\Gamma_0^+ : \Gamma_0] = q_0 + 1, \quad [\Gamma_n : \Gamma_{n-1}] = q_n, \quad \text{for } n \geq 1.$$

The corresponding tree $X = \widetilde{(A, i, 0)}$ admits an action by $\Gamma = \pi_1(\mathbf{A}, 0)$, and we have

(4) $$\Gamma = \Gamma_0^+ *_{\Gamma_0} \Gamma_\infty, \quad \text{where } \Gamma_\infty = \bigcup_{n \geq 0} \Gamma_n.$$

The action of Γ on X is faithful, i.e., $\Gamma \leq G = \operatorname{Aut}(X)$, iff

(5) Γ_0 contains no non-trivial subgroup normal in both Γ_0^+ and Γ_∞.

We shall assume this to be the case.

Further, Γ is then an X-lattice iff the Γ_n are finite, in which case

(6)
$$\mathrm{Vol}(\Gamma\backslash\backslash X) = \frac{1}{|\Gamma_0^+|} + \sum_{n\geq 1}\frac{1}{|\Gamma_n|} = \frac{1}{|\Gamma_0|}\left(\frac{1}{q_0+1} + \sum_{n\geq 1}\frac{1}{q_1\cdots q_n}\right) < \infty \quad \text{(by (2))}.$$

From (2) we see that (A, i), hence also X, has no terminal vertices, and so, by (5.12),

(7) Γ acts minimally on X with non-abelian length function ℓ.

(cf. (5.13), (5.14)). It follows then from (5.4) that

(8) $Z_G(\Gamma) = \{1\}.$

To calculate $N_G(\Gamma)/\Gamma$ we shall use the methods of [BJ], which are applicable thanks to (7). First, since the graph A has no non-trivial automorphisms, we have an epimorphism

(9) $\mathrm{Aut}(\mathbf{A}) \longrightarrow \mathrm{Aut}(\Gamma)_\ell \cong N_G(\Gamma),$

where $\mathbf{A}$ denotes the graph of groups (3), and $\mathrm{Aut}(\Gamma)_\ell$ denotes the stabilizer of ℓ, consisting of all $\alpha \in \mathrm{Aut}(\Gamma)$ such that $\ell \circ \alpha = \ell$. In the notation of [BJ], (9) is the map $\Phi \mapsto \Phi_0$, where the subscript refers to the base point vertex 0 in A ([BJ], Theorem (4.1)). Since we are interested in the image mod $\mathrm{ad}(\Gamma)$, we can restrict attention to "$\delta\mathrm{Aut}(\mathbf{A})$." We now freely use the notation of [BJ] (cf. [BJ](6.3)(6)), and so assume that

(10)
$$\begin{aligned}
\Phi &= \delta\Phi = (\phi, (\delta)), \\
\phi_A &= Id_A, \\
\phi_n &\in \mathrm{Aut}(\Gamma_n) \quad \text{for } n \geq 1, \quad \phi_0 \in \mathrm{Aut}(\Gamma_0^+), \\
\phi_{e_n} &\in \mathrm{Aut}(\Gamma_n) \quad \text{for } n \geq 0, \\
\delta_n(&:= \delta_{e_n}) \in \Gamma_n \quad \text{for } n > 0, \quad \text{and} \\
\delta_0(&= \delta_{e_0}) \in \Gamma_0^+, \quad \text{and} \\
\bar{\delta}_n(&:= \delta_{\bar{e}_n}) \in \Gamma_{n+1} \quad \text{for } n > 0.
\end{aligned}$$

These data are required to yield commutative diagrams

$$
\begin{array}{ccccc}
\Gamma_0^+ & > & \Gamma_0 & < & \Gamma_1 \\
(11)_0 \qquad ad(\delta_0^{-1})\varphi_0 \ \downarrow & & \downarrow \ \varphi_{e_0} & & \downarrow \ ad(\bar{\delta}_0^{-1})\varphi_1 \\
\Gamma_0^+ & > & \Gamma_0 & < & \Gamma_1
\end{array}
$$

and, for $n > 0$,

$$
\begin{array}{ccccc}
\Gamma_n & = & \Gamma_n & < & \Gamma_{n+1} \\
\end{array}
$$

$(11)_n$ $ad(\delta_n^{-1})\varphi_n \downarrow$ $\downarrow \varphi_{e_n}$ $\downarrow ad(\bar{\delta}_n^{-1})\varphi_1$

$$
\begin{array}{ccccc}
\Gamma_n & = & \Gamma_n & < & \Gamma_{n+1}.
\end{array}
$$

Put

(12) $\Delta_0 = \delta_0^{-1}(\in \Gamma_0^+)$ and $\Delta_{n+1} = \Delta_n \delta_n \bar{\delta}_n^{\;-1}(\in \Gamma_{n+1})$ for $n \geq 0$.

Note then that

(13) $\sigma_n := \Delta_n \delta_n$ satisfies $\sigma_0 = 1$ and $\sigma_n \in \Gamma_n$ for $n > 0$.

Put

(14) $\psi_n = ad(\Delta_n)\phi_n \in \mathrm{Aut}(\Gamma_n^{(+, \text{ if } n=0)})$.

Then the commutative disgrams $(11)_n$ can be assembled into a commutative diagram

(15)

$$
\begin{array}{ccccccccccc}
\Gamma_0^+ & > & \Gamma_0 & < & \Gamma_1 & < & \Gamma_2 & < & \Gamma_3 & < & \cdots \\
\psi_0 \downarrow & & \varphi_{e_0} \downarrow & & \psi_1 \downarrow & & \psi_2 \downarrow & & \psi_3 \downarrow & & \cdots \\
\Gamma_0^+ & > & \Gamma_0 & < & \Gamma_1 & < & \Gamma_2 & < & \Gamma_3 & < & \cdots
\end{array}
$$

This defines

(16)
$$
\Psi = (\psi, (1)) \in \delta\mathrm{Aut}(\mathbf{A}),
$$
ψ_n as above, $\psi_{e_0} = \phi_{e_0}$, $\psi_{e_n} = \psi_n$ for $n > 0$.

Further define

(17)
$$
\Phi' = (\phi', (\delta')), \quad \delta'_n = \delta_n,
$$
$$
\phi'_n = ad(\Delta_n^{-1}), \quad \phi'_{e_n} = ad(\delta_n^{-1})\phi'_n = ad(\sigma_n^{-1}).
$$

Then $\Phi' \in \delta\mathrm{Aut}(\mathbf{A})$ also, as is easily verified. We further claim that

(18) $\Phi = \Phi' \circ \Psi$.

In fact, put $\Phi'' = \Phi' \circ \Psi = (\phi'', (\delta''))$. Then (cf. [BJ],(3.6))

$$
\phi''_n = \phi'_n \psi_n = ad(\Delta_n^{-1})ad(\Delta_n)\phi_n = \phi_n;
$$
$$
\phi''_{e_n} = \phi'_{e_n}\psi_{e_n} = ad(\delta_n^{-1}\Delta_n^{-1})ad(\Delta_n)\phi_n = ad(\delta_n^{-1})\phi_n = \phi_{e_n} \quad ((11));
$$
$$
\delta''_n = \phi'_n(1)\delta'_n = \delta_n; \quad \text{and}
$$
$$
\bar{\delta}''_n = \phi_{n+1}(1)\bar{\delta}'_n = \bar{\delta}_n,
$$

whence (18).

Consider the path group homomorphism

$$\Phi' : \pi(\mathbf{A}) \longrightarrow \pi(\mathbf{A}).$$

We claim that:

(19)   The automorphism Φ'_0 that Φ' induces on $\Gamma = \pi_1(\mathbf{A}, 0) \leq \pi(\mathbf{A})$ is $ad(\delta_0)$.

Elements of Γ are of the form

$$|\gamma| = g_{n_0} e_{n_1} g_{n_1} e_{n_2} \cdots e_{n_r} g_{n_r},$$

where $\gamma = (g_{n_0}, e_{n_1}, g_{n_1}, e_{n_2}, \cdots, e_{n_r}, g_{n_r})$ is a closed path at the base vertex 0 in $\mathbf{A}$. A segment $(g_{n_{i-1}}, e_{n_i}, g_{n_i})$ of γ is of the form (g, e_n, h) (or its inverse $(h^{-1}, \bar{e}_n, g^{-1})$), where $g \in \Gamma_n$ (or Γ_n^+ if $n = 0$) and $h \in \Gamma_{n+1}$. Now $\Phi'|\Gamma_n = ad(\Delta_n^{-1})$ and $\Phi'(e_n) = \delta_n e_n \bar{\delta}_n^{-1}$ (cf. [BJ]). Hence,

(20) $\Phi'(g e_n h) = (\Delta_n^{-1} g \Delta_n)(\delta_n e_n \bar{\delta}_n^{-1})(\Delta_{n+1}^{-1} h \Delta_{n+1}) = \Delta_n^{-1} g (\sigma_n e_n \sigma_n^{-1}) h \Delta_{n+1}$

(see (13)). From the defining relations for $\pi(\mathbf{A})$ we have

(21) $e_n \alpha_{\bar{e}_n}(s) e_n^{-1} = \alpha_{e_n}(s), \quad \forall s \in \Gamma_{e_n}.$

In $\mathbf{A}$ (see (3)), $\alpha_{e_n} = Id : \Gamma_{e_n} = \Gamma_n \to \Gamma_n$ for $n > 0$, $\alpha_{e_0} : \Gamma_{e_0} = \Gamma_0 \to \Gamma_0^+$ is the inclusion, and $\alpha_{\bar{e}_n} : \Gamma_{e_n} = \Gamma_n \to \Gamma_{n+1}$ is the inclusion for all $n \geq 0$. Thus, in our setting, (21) says that

$$e_n \text{ centralizes } \Gamma_n \text{ in } \pi(\mathbf{A}), \text{ for all } n \geq 0.$$

By (13), $\sigma_n \in \Gamma_n$ for all n. Hence from (20) and (22), we have

$$\Phi'(g e_n h) = \Delta_n^{-1} g e_n h \Delta_{n+1}.$$

It follows directly that if γ is a path in $\mathbf{A}$ from vertex p to vertex q then $\Phi'(|\gamma|) = \Delta_p^{-1} |\gamma| \Delta_q$. Taking $p = q = 0$ we find that Φ' on $\Gamma = \pi_1(\mathbf{A}, 0)$ is $ad(\Delta_0^{-1}) = ad(\delta_0)$ (cf. (12)), whence (19).

Since we are interested in the image of $\delta \mathrm{Aut}(\mathbf{A}) \to \mathrm{Aut}(\Gamma)$ ($\Phi \mapsto \Phi_0$) only mod $ad(\Gamma)$, and since $\Phi = \Phi' \Psi(18)$, with $\Phi'_0 = ad(\delta_0)$ (19), we may replace Φ by Ψ. Let us change notation in Ψ slightly, denoting ψ_0 by ψ_0^+, and $\psi_{e_0} = \phi_{e_0}$ by ψ_0. Then the commutative diagram (15) takes the form

$$
\begin{array}{ccccccccccc}
\Gamma_0^+ & > & \Gamma_0 & < & \Gamma_1 & < & \Gamma_2 & < & \Gamma_3 & < & \cdots \\
\psi_0^+ \downarrow & & \psi_0 \downarrow & & \psi_1 \downarrow & & \psi_2 \downarrow & & \psi_3 \downarrow & & \cdots \\
\Gamma_0^+ & > & \Gamma_0 & < & \Gamma_1 & < & \Gamma_2 & < & \Gamma_3 & < & \cdots
\end{array}
$$

(23)

The ψ_n $(n \geq 0)$ assemble to give

$$\psi_\infty \in \text{Aut}^{\text{filt}}(\Gamma_\infty), \qquad \Gamma_\infty = \bigcup_{n \geq 0} \Gamma_n,$$

and then Ψ_0 is the amalgam automorphism

(24) $$\Psi_0 = \psi_0^+ *_{\psi_0} \psi_\infty \in \text{Aut}(\Gamma = \Gamma_0^+ *_{\Gamma_0} \Gamma_\infty),$$

Now the filtered amalgam automorphisms of Γ, i.e., those preserving Γ_0^+ and Γ_∞, as well as the filtration of Γ_∞, form a group naturally isomorphic to the fiber product,

(25) $$\text{Aut}(\Gamma_0^+; \Gamma_0) \times_{\text{Aut}(\Gamma_0)} \text{Aut}(\Gamma_\infty; (\Gamma_n))$$

relative to the restriction homomorphism from the two factors to $\text{Aut}(\Gamma_0)$. In a nontrivial amalgam $H = U *_V W$, with V a proper subgroup of U and of W, one has $N_H(U) \cap N_H(W) = U \cap W = V$. Hence, if an amalgam automorphism α of H is inner, $\alpha = ad(h)$, we must have $h \in V$.

It follows that, in $\text{Aut}(\Gamma)$, the intersection of (25) with $ad(\Gamma)$ is $ad_\Gamma(\Gamma_0)$. The discussion above shows that (25) maps into $N_G(\Gamma)$ mod $ad(\Gamma)$. Thus we have shown that

(26) $$N_G(\Gamma)/\Gamma \cong [\text{Aut}(\Gamma_0^+; \Gamma) \times_{\text{Aut}(\Gamma_0)} \text{Aut}(\Gamma_\infty; (\Gamma_n)_{n \geq 0})]/ad_\Gamma(\Gamma_0),$$

where the numerator denotes the group of pairs $\psi = (\psi_0^+, \psi_\infty) \in \text{Aut}(\Gamma_0^+) \times \text{Aut}(\Gamma_\infty)$ such that $\psi_0^+ \Gamma_0 = \Gamma_0$, $\psi_\infty \Gamma_n = \Gamma_n$ for all $n \geq 0$, and $\psi_0^+|\Gamma_0 = \psi_\infty|\Gamma_0$, and the denominator consists of pairs $(ad_{\Gamma_0^+}(g), ad_{\Gamma_\infty}(g))$, $g \in \Gamma_0$.

We next investigate the finite and locally finite subgroups of $\Gamma = \Gamma_0^+ *_{\Gamma_0} \Gamma_\infty$. In the tree X, let ∞ denote the end fixed by Γ_∞, and

(27) $$[x_0, \infty) = \quad \underset{x_0}{\circ} \overset{e_0}{\underline{\hspace{1.2cm}}} \underset{x_1}{\circ} \overset{e_1}{\underline{\hspace{1.2cm}}} \underset{x_2}{\circ} \overset{e_2}{\underline{\hspace{1.2cm}}} \underset{x_3}{\circ} \overset{e_3}{\underline{\hspace{1.2cm}}} \underset{x_4}{\circ} \underline{\hspace{1cm}} \cdots$$

the fundamental domain for Γ on X such that

(28) $$\begin{aligned} \Gamma_{e_n} &= \Gamma_n \quad (n \geq 0) \\ &= \Gamma_{x_n} \quad (n \geq 1), \quad \text{and} \\ \Gamma_{x_0} &= \Gamma_0^+. \end{aligned}$$

We claim the following:

(29) Let $F \leq \Gamma = \Gamma_0^+ *_{\Gamma_0} \Gamma_\infty$ be a finite subgroup not Γ-conjugate to a subgroup of Γ_0. Let X^F denote the tree of fixed points of F. Then exactly one of the following occurs.

(a) $X^F = \{gx_0\}$ for some $g \in \Gamma$, and $F \leq g\Gamma_0^+ g^{-1}$, or

(b) F fixes a unique end, ε_F, $\varepsilon_F = g\infty$ for some $g \in \Gamma$, and $F \leq g\Gamma_\infty g^{-1}$.

We begin with a general remark about amalgams.

(30)
$$\text{In an amalgam } D = A *_C B, \quad \text{for } s \in D - B,$$
$$B \cap s B s^{-1} = B \cap t C t^{-1} \quad \text{for some } t \in D.$$

In fact, write s as a reduced word in $A *_C B$, and let t be the result of removing the right-hand factor, in case that factor belongs to B, and $t = s$ otherwise. Then $s B s^{-1} = t B t^{-1}$, and t still has length ≥ 1, since $s \notin B$. Let $b \in B$ be so that $t b t^{-1} \in B$ also. If $b \notin C$ then $t b t^{-1}$ is a reduced word, and of length ≥ 3, hence can not belong to B. Thus $b \in C$, thus proving (30).

Now to prove (29) we first note that F is Γ-*conjugate* to a subgroup of Γ_0^+ or of Γ_∞ (cf. [Se], I, Th. 8). Say $F \leq g \Gamma_0^+ g^{-1} = \Gamma_{g x_0}$. Since $\Gamma_{x_0} = \Gamma_0^+$ acts transitively on $E_0(x_0)$, and $\Gamma_{e_0} = \Gamma_0$, our hypothesis that F is not conjugate to a subgroup of Γ_0 implies that F fixes no edge of $E_0(g x_0)$, whence $X^F = \{g x_0\}$.

Next consider the case $F \leq g \Gamma_\infty g^{-1} = \Gamma_{g \infty}$. We must show that $g \infty$ is the unique end of X^F. After conjugation by g^{-1} we can assume that $g = 1$, $F \leq \Gamma_\infty$.

We have $F \leq \Gamma_{n_0}$ for some n_0, but $F \nleq \Gamma_0$, by hypothesis. Let $x \in V X^F \subset V X = \coprod_{n \geq 0} \Gamma \cdot x_n$, say $x = s x_n$ $(s \in \Gamma, n \geq 0)$. Then $F \leq \Gamma_{s x_n} = s \Gamma_{x_n} s^{-1}$. If $n = 0$, then $F \leq s \Gamma_0^+ s^{-1} \cap \Gamma_\infty \leq t \Gamma_0 t^{-1} \cap \Gamma_\infty$ for some $t \in \Gamma$ (by (30)), contrary to assumption. Thus $n \geq 1$ and $F \leq s \Gamma_n s^{-1} \cap \Gamma_\infty \leq s \Gamma_\infty s^{-1} \cap \Gamma_\infty$. Since, by assumption, $F \nleq s \Gamma_0 s^{-1} \cap \Gamma_\infty$, it follows from (30) that $s \in \Gamma_\infty$, and so $x = s x_n \in \Gamma_\infty x_n$. Thus

$$X^F \subset \Gamma_\infty \cdot [x_1, \infty).$$

Now it is easily seen that $\Gamma_\infty \cdot [x_1, \infty)$, is a parabolic tree, with unique end ∞. This proves (29).

(30)   If $F \leq \Gamma$ is an infinite locally finite subgroup then F fixes a unique end ε_F of
       X, $\varepsilon_F = g \infty$ for some $g \in \Gamma$, and $F \leq g \Gamma_\infty g^{-1}$.

In fact if $F' \leq F$ is a finite subgroup of order $|F'| > |\Gamma_0^+|$ then F' fixes a unique end $\varepsilon_{F'} = g_{F'} \infty$, by (29)(b). If F'' is a larger finite subgroup then, by uniqueness, $\varepsilon_{F''} = \varepsilon_{F'}$. Since F is the directed union of such finite subgroups F', the result follows.

This has the following consequence for the commensurator $C_G(\Gamma)$.

(32)
$$C_G(\Gamma) = \Gamma \cdot C_G(\Gamma)_\infty, \quad \text{and}$$
$$C_G(\Gamma)_\infty \leq C_G(\Gamma_\infty).$$

In fact, let $g \in C_G(\Gamma)$. Then $F = g \Gamma g^{-1} \cap \Gamma_\infty$ has finite index in Γ_∞, hence fixes a unique end $h \infty$ $(h \in \Gamma)$ of X. Thus $g = h(h^{-1} g)$ with $h^{-1} g \in C_G(\Gamma)_\infty$.

Suppose that $g \in C_G(\Gamma)_\infty$. Then Γ_∞ contains with finite index the subgroup $\Gamma_\infty \cap g \Gamma g^{-1} = \Gamma_\infty \cap (g \Gamma g^{-1})_\infty = \Gamma_\infty \cap g \Gamma_{g^{-1} \infty} g^{-1} = \Gamma_\infty \cap g \Gamma_\infty g^{-1}$, and similarly for g^{-1}. Thus $C_G(\Gamma)_\infty \leq C_G(\Gamma_\infty)$.

Finally, we observe that:

(33)   (a) If $q_n = q$ for all $n \geq 0$, then $X = X_{q+1}$, the $(q+1)$-regular tree.
       (b) We have $G\backslash X = A$ unless, for some $r > 0$, we have, for all $n \geq 0$,

$$q_{2nr} = q_0, \quad q_{(2n+1)r} = q_1 \geq 2, \quad \text{and} \quad q_n = 1 \quad \text{if } r \nmid n.$$

In this case,

$$I(G\backslash\backslash X) = \underset{q_0+1}{\circ}\!\!\overset{\quad 1 \quad 1}{\underline{\hspace{1.3cm}}}\!\underset{}{\circ}\!\overset{\quad 1}{\underline{\hspace{1cm}}}\!\underset{}{\circ}\ \cdots\ \overset{1 \quad 1}{\underline{\hspace{1cm}}}\!\underset{}{\circ}\!\overset{1 \quad 1}{\underline{\hspace{1.3cm}}}\!\underset{q_r+1}{\circ}$$

and X is a subdivision of the bi-regular tree, X_{q_0+1,q_r+1}.
       (c) X has bounded degree iff the q_n are bounded.

Assertions (a) and (c) are immediate. For assertion (b) consider the covering of indexed graphs, $p : (A, i) \to (B, j) = I(G\backslash\backslash X)$. For $a \in VA$, $deg_B(p(a)) \leq deg_A(a) \leq 2$. It follows easily from this that, if p is not an isomorphism, then it must fold A back and forth onto the finite interval B, and then the indexed-covering property $((2.5)(5))$ gives the indicated conditions on the indices q_n.

Consider the *non-uniform case* of (33)(b). Then the projection $p : X \to A = \Gamma\backslash X = G\backslash X$ defines "levels" $p : VX \to \mathbf{N}$ on VX which are preserved by G. It follows that, along any hyperbolic axis in X, the levels must be periodic, hence bounded. If therefore ε is an end of X of "unbounded level," i.e., a ray toward ε traverses unbounded levels, then $G_\varepsilon = G_\varepsilon^0$, i.e., G_ε contains no hyperbolic elements.

With the notation of (27), consider G_∞. If $g \in G_\infty$ then, by the previous paragraph, g fixes all points of $[x_n, \infty)$ for some $n \geq 0$. If $n \geq 0$, then Γ_n acts transitively on $E_0(x_n) - \{e_n\}$. Modifying g by an element of Γ_n we can therefore arrange that g also fixes e_{n-1}, and $x_{n-1} = \partial_0 e_{n-1}$. Thus we prove inductively that

(34) $G_\infty = \Gamma_\infty \cdot G_{[x_0,\infty)}$,

where $G_{[x_0,\infty)} = G_{x_0} \cap G_\infty$. Combining (34) with (32), we have:

(35)   In the non-uniform case of (33)(b), we have $C_G(\Gamma) = \Gamma \cdot C_G(\Gamma)_{[x_0,\infty)}$.

10.2   Nagao's Theorem: $\Gamma = PGL_2(\mathbf{F}_q[t])$

For a commutative ring A we have

(1) $PGL_2(A) = GL_2(A)/A^\times \cdot I.$

For $a, d \in A^\times$ and $b \in A$ we have

(2) $h(a, d) = \begin{pmatrix} a & 0 \\ 0 & d \end{pmatrix}, \quad u(b) = \begin{pmatrix} 1 & b \\ 0 & 1 \end{pmatrix} \quad \text{in } PGL_2(A)$

with $h(a, d) = I$ iff $a = d$, and

(3) $$h(a, d)u(b)h(a, d)^{-1} = u(abd^{-1}).$$

Thus

$$T(A) := \{h(a, d) \mid a, d \in A^\times\} \cong A^\times,$$

(4) $$U(A) \cong A, \quad \text{and}$$

$$B(A) := \left\{ \begin{pmatrix} a & b \\ 0 & d \end{pmatrix} \in PGL_2(A) \right\} = T(A) \bowtie U(A) \cong A^\times \bowtie A.$$

Now let $A = \mathbf{F}_q[t]$, a polynomial algebra over a field of cardinal q. Then $A^\times = \mathbf{F}_q^\times$. It follows from [Se] (Ch. II, (1.6)) that

$$\Gamma := PGL_2(A)$$

(5) $$= PGL_2(\mathbf{F}_q) *_{B(\mathbf{F}_q)} B(A)$$

$$= \Gamma_0^+ *_{\Gamma_0} \Gamma_\infty, \quad \Gamma_\infty = \cup_{n \geq 0} \Gamma_n,$$

where $\Gamma_0^+ = PGL_2(\mathbf{F}_q)$ and, for $n \geq 0$,

$$\Gamma_n = \left\{ \begin{pmatrix} a & b \\ 0 & d \end{pmatrix} \in \Gamma \mid deg_t(b) \leq n \right\}$$

This defines a grouping of

(6)
$$\underset{\circ}{\overset{q+1}{\rule{3.5cm}{0.4pt}}} \underset{\circ}{\overset{q}{\rule{1cm}{0.4pt}}} \overset{1}{\underset{\circ}{\rule{1cm}{0.4pt}}} \overset{q}{\rule{1cm}{0.4pt}} \overset{1}{\underset{\circ}{\rule{1cm}{0.4pt}}} \overset{q}{\rule{1cm}{0.4pt}} \overset{1}{\rule{1cm}{0.4pt}} \cdots$$

with universal cover $X = X_{q+1}$ the $(q + 1)$-regular tree. In fact, putting

(7) $$F = \mathbf{F}_q(t) \quad \text{and} \quad \hat{F} = \mathbf{F}_q((t^{-1})),$$

we have

(8) $\Gamma = PGL_2(\mathbf{F}_q[t]) \leq H = PGL_2(F) \leq \bar{H} := PGL_2(\hat{F}) \leq G = \text{Aut}(X),$

and X is the Bruhat–Tits tree of $\bar{H}$. Further Γ is a non-uniform $\bar{H}$-lattice, an X-lattice with edge-indexed quotient (6), and ((10.1)(6))

(9) $$\text{Vol}(\Gamma \backslash\backslash X) = \frac{1}{(q-1)q} \left(\frac{1}{q+1} + \sum_{n \geq 1} \frac{1}{q^n} \right) = \frac{2}{(q-1)(q^2-1)}.$$

Moreover ((10.1)(8)),

(10) $$Z_G(\Gamma) = \{1\}.$$

It is known that

(11) $$N_{\bar{H}}(\Gamma) = \Gamma \quad \text{and} \quad C_{\bar{H}}(\Gamma) = H.$$

To calculate $N_G(\Gamma)/\Gamma$ we use (10.1)(26). First observe that $\Gamma_0^+ (= PGL_2(\mathbf{F}_q))$ is inn-exact (cf. (6.1)), and $N_{\Gamma_0^+}(\Gamma_0) = \Gamma_0$, whence an isomorphism

$$ad_{\Gamma_0^+} : \Gamma_0 \xrightarrow{\cong} \text{Aut}(\Gamma_0^+; \Gamma_0).$$

It follows therefore from (10.1)(26) that

(12) $$\begin{aligned} N_G(\Gamma)/\Gamma &\cong \text{Aut}_0^{\text{filt}}(\Gamma_\infty)/ad_{\Gamma_\infty}(\Gamma_0), \quad \text{where} \\ \text{Aut}_0^{\text{filt}}(\Gamma_\infty) &= \{\alpha \in \text{Aut}^{\text{filt}}(\Gamma_\infty) \mid \alpha|\Gamma_0 \in ad(\Gamma_0)\} \end{aligned}.$$

We have a split exact sequence

(13) $$1 \to \Gamma_\infty^t \to \Gamma_\infty \to \Gamma_0 \to 1$$

$$\|$$

$$U(tA) = \begin{bmatrix} 1 & t\mathbf{F}_q[t] \\ 0 & 1 \end{bmatrix}$$

and an exact sequence

(14) $$1 \to \text{Aut}_t^{\text{filt}}(\Gamma_\infty) \to \text{Aut}_0^{\text{filt}}(\Gamma_\infty) \xrightarrow{\text{res}} ad(\Gamma_0) \to 1$$

$$\wedge$$

$$\text{Aut}(\Gamma_0)$$

split by $ad_{\Gamma_\infty} : \Gamma_0 \to \text{Aut}_0^{\text{filt}}(\Gamma_\infty)$. Thus,

(15) $$\text{Aut}_0^{\text{filt}}(\Gamma_\infty) = \text{Aut}_t^{\text{filt}}(\Gamma_\infty) \rtimes ad_{\Gamma_\infty}(\Gamma_0),$$

whence, from (12),

(16) $$N_G(\Gamma)/\Gamma \cong \text{Aut}_t^{\text{filt}}(\Gamma_\infty) = \{\alpha \in \text{Aut}^{\text{filt}}(\Gamma_\infty) \mid \alpha|\Gamma_0 = Id\}$$

From the isomorphism (4) $\Gamma_\infty = B(A) \cong \mathbf{F}_q^\times \ltimes \mathbf{F}_q[t]$, with $\Gamma_0 = \mathbf{F}_q^\times \ltimes \mathbf{F}_q$, we see that $N_G(\Gamma)/\Gamma$ in (16) is isomorphic to

$$\text{Aut}_t^{\text{filt}}(A) = \{\alpha \in \text{Aut}_{\mathbf{F}_q}(A) \mid \alpha(1) = 1 \text{ and } \deg_t(\alpha(t^n)) = n, \forall n \geq 0\}$$

(17) $$\cong \begin{bmatrix} 1 & * & * & * & \\ & * & * & * & \\ & & * & * & * \\ 0 & & & * & \\ & & & & \ddots \end{bmatrix} \leq GL_\infty(\mathbf{F}_q).$$

Contrast (16) and (17) with (11): $N_{\bar{H}}(\Gamma) = \Gamma$.

Shahar Mozes [M2] has proved that $C_G(\Gamma)$ is dense in G.

10.3   A divisible $(q + 1)$-regular grouping

Consider the grouping

(1)

$$\overset{\Gamma_0^+ \qquad\qquad \Gamma_1 \qquad\qquad \Gamma_2 \qquad\qquad \Gamma_3}{\underset{\Gamma_0 \qquad\qquad \Gamma_1 \qquad\qquad \Gamma_2 \qquad\qquad \Gamma_3}{\circ\!\!-\!\!-\!\!-\!\!-\!\!\circ\!\!-\!\!-\!\!-\!\!-\!\!\circ\!\!-\!\!-\!\!-\!\!-\!\!\circ\!\!-\!\!-\!\!-\!\!-}} \cdots$$

of

(2)

$$\underset{q+1 \qquad\quad q \quad 1 \qquad q \quad 1 \qquad q \quad 1}{\circ\!\!-\!\!-\!\!-\!\!-\!\!\circ\!\!-\!\!\circ\!\!-\!\!-\!\!\circ\!\!-\!\!\circ\!\!-\!\!-\!\!\circ\!\!-\!\!\circ} \cdots$$

in (10.2). Up to isomorphism (cf. (10.2)(4) and (5)),

(3)
$$\Gamma_n \cong \mathbf{F}_q^\times \ltimes A_n, \quad \text{where}$$
$$A_n = \mathbf{F}_q \cdot 1 + \mathbf{F}_q t + \cdots + \mathbf{F}_q t^n \le A = \mathbf{F}_q[t].$$

We propose now to replace A_n by a group B_n, defined as follows. Let E be the unramified extension of $\mathbf{Q}_p$ $(p = char(\mathbf{F}_q))$ with residue field

(4)
$$\mathbf{F}_q = R/pR, \quad R = \text{the integers of } E.$$

Then put

(5)
$$B_n = p^{-(n+1)} R/R \le E/R = B_\infty.$$

The multiplicative lift $\mathbf{F}_q^\times \to R^\times \le E^\times$ defines an action of $\mathbf{F}_q$ on E, hence also on B_∞, stabilizing each B_n. We put

(6)
$$\Gamma_n' = \mathbf{F}_q^\times \ltimes B_n$$

and so obtain a new grouping $\mathbf{A}$ of (2),

(7)

$$\overset{\Gamma_0^+ \qquad\qquad \Gamma_1' \qquad\qquad \Gamma_2' \qquad\qquad \Gamma_3'}{\underset{\Gamma_0 \cong \Gamma_0' \qquad \Gamma_1' \qquad\qquad \Gamma_2' \qquad\qquad \Gamma_3'}{\circ\!\!-\!\!-\!\!-\!\!-\!\!\circ\!\!-\!\!-\!\!-\!\!-\!\!\circ\!\!-\!\!-\!\!-\!\!-\!\!\circ\!\!-\!\!-\!\!-\!\!-}} \cdots$$

As in (10.1) we see that, with $\Gamma' = \pi_1(\mathbf{A}', 0)) = \Gamma_0^+ *_{\Gamma_0} \Gamma_\infty'$

(8)
$$Z_G(\Gamma') = \{1\}, \qquad \text{Vol}(\Gamma' \backslash\backslash X) = \text{Vol}(\Gamma \backslash\backslash X),$$

and

(9)
$$N_G(\Gamma')/\Gamma' \cong \text{Aut}_t^{\text{filt}}(\Gamma_\infty') = \{\alpha \in \text{Aut}_{\mathbf{F}_q}(B_\infty) \mid \alpha|B_0 = Id\}.$$

It is easily seen that $End_{\mathbf{F}_q^\times}(B_\infty) \cong R$, and restriction to B_0 corresponds to reduction *mod pR*, whence

(10) $$N_G(\Gamma')/\Gamma' \cong R_1^\times = Ker(R^\times \to \mathbf{F}_q^\times) = 1 + pR,$$

a *finitely generated abelian pro-p group*. Since B_∞ is divisible, it lies in $P(\Gamma') = Ker(\Gamma' \to \hat{\Gamma}')$. The normal subgroup of Γ_0^+ generated by B_0 is all of Γ_0^+, hence $P(\Gamma')$ contains Γ_0^+ and also $\Gamma'_\infty = \Gamma'_0 \cdot B_\infty$, whence

(11) $$P(\Gamma') = \Gamma', \quad \text{i.e., } \hat{\Gamma}' = \{1\}.$$

It follows therefore from (6.1)(14) that

(12) $$C_G(\Gamma') = N_G(\Gamma').$$

Since $G = \mathrm{Aut}(X_{q+1})$ has a simple subgroup of index 2 (see (6.12)), Γ' cannot be normal in G, so

(13) $$C_G(\Gamma') \text{ is closed, and not dense in } G.$$

10.4   The PNeumann groupings

We shall make use of the PNuemann groups constructed in Appendix [PN]. Let S be a simple inn-exact group of order m, and V a transitive rigid S-set of size $r > 1$. (See Appendix [PN], (1.0), for terminology.) Put

(1) $$q = m^{r^2+1}.$$

Then we have groups

(2) $$\Gamma_0 < \Gamma_1 < \Gamma_2 < \cdots ; \quad \Gamma_\infty = \cup_{n \geq 0}\Gamma_n,$$

with the following properties, for all $n \geq 0$.

(3) $$|\Gamma_0| = m^2 \quad \text{and} \quad [\Gamma_{n+1} : \Gamma_n] = q,$$

(4) $$ad_{\Gamma_{n+1}} : \Gamma_n \xrightarrow{\cong} \mathrm{Aut}(\Gamma_{n+1}; \Gamma_n), \quad \text{and}$$
$$ad_{\Gamma_\infty} : \Gamma_0 \xrightarrow{\cong} \mathrm{Aut}^{\mathrm{filt}}(\Gamma_\infty).$$

(5) Γ_0 contains no non-trivial normal subgroup of Γ_∞.

Choose a finite group Γ_0^+ containing Γ_0, and put

(6) $$q_0 + 1 = [\Gamma_0^+ : \Gamma_0].$$

Then we have the faithful grouping.

(7)

$$
\begin{array}{ccccc}
\Gamma_0^+ & \Gamma_1 & \Gamma_2 & \Gamma_3 & \\
\circ\!\!-\!\!\!-\!\!\!-\!\!\!-\!\!\!\circ\!\!-\!\!\!-\!\!\!-\!\!\!-\!\!\!\circ\!\!-\!\!\!-\!\!\!-\!\!\!-\!\!\!\circ\!\!-\!\!\!-\!\!\!- & \cdots \\
\Gamma_0 & \Gamma_1 & \Gamma_2 & \Gamma_3 &
\end{array}
$$

of

(8)

$$
\begin{array}{ccccccc}
& q_0+1 & & q & 1 & q & 1 & q & 1 \\
\cup\!\!-\!\!\!-\!\!\!-\!\!\!-\!\!\!-\!\!\!\circ\!\!-\!\!\!-\!\!\!-\!\!\!\circ\!\!-\!\!\!-\!\!\!-\!\!\!\circ\!\!-\!\!\!-\!\!\!- & \cdots
\end{array}
$$

with volume

(9) $$\mathrm{Vol}(\Gamma\backslash\backslash X) = \frac{1}{m^2}\left(\frac{1}{q_0+1} + \sum_{n\geq 1}\frac{1}{q^n}\right) = \frac{q_0+q}{m^2(q_0+1)(q-1)}.$$

As always ((10.1)(8))

(10) $$Z_G(\Gamma) = \{1\}.$$

From (10.1)(26) and (4) above,

(11)
$$N_G(\Gamma)/\Gamma \cong \mathrm{Aut}_0(\Gamma_0^+;\Gamma_0)/ad_{\Gamma_0^+}(\Gamma_0), \quad \text{where}$$
$$\mathrm{Aut}_0(\Gamma_0^+;\Gamma_0) = \{\alpha \in \mathrm{Aut}(\Gamma_0^+;\Gamma_0)|\ \alpha|\Gamma_0 \in ad(\Gamma_0)\}.$$

It follows that

(12)
$$
\begin{aligned}
N_G(\Gamma)/\Gamma &\cong \mathrm{Aut}_1(\Gamma_0^+;\Gamma_0) \\
&= Ker(\mathrm{Aut}_0(\Gamma_0^+;\Gamma_0) \xrightarrow{\ \mathrm{res}\ } ad(\Gamma_0)) \\
&= \{\alpha \in \mathrm{Aut}(\Gamma_0^+)|\ \alpha|\Gamma_0 = Id\},
\end{aligned}
$$

a finite group.

  We mention two kinds of choices for Γ_0^+. First,

(13)   Taking $\Gamma_0^+ = \Gamma_1$, we have $q_0+1 = q$, $\mathrm{Vol}(\Gamma\backslash\backslash X) = \frac{2q-1}{m^2 q(q-1)}$, and $N_G(\Gamma)$
     $= \Gamma$.

On the other hand,

(14)   Taking $\Gamma_0^+ = Q \times \Gamma_0$, with Q any group of order $q+1$, we have $q_0 = q$, $X =$
     X_{q+1}, the $(q+1)$-regular tree, $\Gamma = (Q \times \Gamma_0) *_{\Gamma_0} \Gamma_\infty = (Q * \Gamma_\infty)/(Q,\Gamma_0)$,
     $\mathrm{Vol}(\Gamma\backslash\backslash X) = \frac{2q}{m^2(q^2-1)} = \frac{2m^{r^2-1}}{m^{2r^2+2}-1}$, and $N_G(\Gamma) \cong \Gamma \rtimes \mathrm{Aut}(Q)$, where $\mathrm{Aut}(Q)$
     acts trivially on $\Gamma_\infty \leq \Gamma$.

  Put $H = \mathrm{Aut}(Q)$ and $N = N_G(\Gamma) \cong [(Q * \Gamma_\infty)/(Q,\Gamma_0)]\rtimes H$. Then, since H
normalizes Q and Γ_∞, it fixes the unique fixed point, x_0 of Q, and the unique fixed

end, ∞, of Γ_∞. It follows easily that $\Gamma \backslash X \to N \backslash X$ is an isomorphism, so N has the same fundamental domain as Γ, with stabilizers:

$$N_{x_0} = \Gamma_0^+ \rtimes H = \Gamma_0 \times (Q \rtimes H),$$
$$N_{x_n} = \Gamma_n \times H \text{ for } n \geq 1; \quad N_\infty = \Gamma_\infty \times H.$$

Taking $Q = \mathbf{Z}/(q+1)\mathbf{Z}$, cyclic, we have $H = Q^\times$, abelian. Since each Γ_n has simple Jordan–Holder factors, $\Gamma_n = (\Gamma_n, \Gamma_n)$ for $n \leq \infty$, so that the normal closure R of Γ_∞ in Γ is characteristic in Γ, hence characteristic in N, since $N/R \cong Q \rtimes Q^\times$ is solvable. Thus, $mod\ ad(\Gamma)$, each element $\alpha \in N_G(N)$ stabilizes Γ_∞ : $\alpha(\Gamma_\infty) = \Gamma_\infty$. Then $Z_N(\Gamma_\infty) = H$ is also α-invariant, and $\alpha(\infty) = \infty$, so $\alpha(x_n) = x_n$ for $n \gg 0$.

We claim that $\alpha(Q) \leq \Gamma$. In fact, passing to $N/R = Q \rtimes Q^\times$, it suffices to see that the automorphism induced by α leaves Q invariant. For this it suffices to see that Q is characteristic in $Q \rtimes Q^\times$. In fact Q is the commutator subgroup of $Q \times Q^\times$. (This follows since $q + 1 \doteq m^{n^2+1} + 1$ is odd, m being the order of a simple non-abelian group. If $R = \mathbf{Z}/N\mathbf{Z}$, the image of R in $(R \rtimes R^\times)^{ab}$ is R/J, where J is the ideal generated by all $1 - u$ ($u \in R^\times$), i.e., the largest quotient of R with trivial unit group. Thus $R/J = \mathbf{Z}/M\mathbf{Z}$ with $M|N$ and $\varphi(M) = 1$, i.e., $M = 1$ or 2. If N is odd, $R/J = \{0\}$.)

The above discussion shows that $\Gamma \lhd N_G(N)$, whence $N_G(N) = N$ (because $Z_G(\Gamma) = \{1\}$). Conclusion:

(15)   Taking Q cyclic in (14), $N_G(\Gamma)$ is self-normalizing in G.

10.5   The symmetric groupings

Fix an integer $d \geq 2$. For $n \geq 0$ put

(1)
$$\Gamma_n = S_{d+n} \quad \text{(symmetric group), and}$$
$$\Gamma'_n = A_{d+n} \quad \text{(alternating group)}.$$

Also put

(2) $\Gamma_0^+ = \Gamma_1, \qquad \Gamma_0'^+ = \Gamma_1'.$

Then we have faithful groupings

(3)

(4) $\underset{\circ}{\overset{d+1}{\rule{0pt}{0pt}}} \rule[0.3em]{6em}{0.4pt} \underset{\circ}{\overset{d+1}{\rule{0pt}{0pt}}} \overset{1}{\rule[0.3em]{4em}{0.4pt}} \underset{\circ}{\overset{d+2}{\rule{0pt}{0pt}}} \overset{1}{\rule[0.3em]{4em}{0.4pt}} \underset{\circ}{\overset{d+3}{\rule{0pt}{0pt}}} \overset{1}{\rule[0.3em]{4em}{0.4pt}} \quad \cdots$

The corresponding X-lattices are

$$
\begin{aligned}
\Gamma &= \Gamma_0^+ *_{\Gamma_0} \Gamma_\infty = S_{d+1} *_{S_d} S_\infty, \quad \text{and} \\
\Gamma' &= \Gamma_0'^+ *_{\Gamma_0'} \Gamma_\infty' = A_{d+1} *_{A_d} A_\infty, \quad \text{with} \\
\Gamma' &\leq \Gamma, \quad [\Gamma : \Gamma'] = 2.
\end{aligned}
\tag{5}
$$

We have

$$
\begin{aligned}
\mathrm{Vol}(\Gamma \backslash\backslash X) &= \frac{1}{(d+1)!} + \sum_{m>d} \frac{1}{m!} \\
&= e - \left(1 + \frac{1}{2!} + \cdots + \frac{1}{d!}\right) + \frac{1}{(d+1)!} \\
&= \frac{1}{2} Vol(\Gamma' \backslash\backslash X).
\end{aligned}
\tag{6}
$$

Further (cf. (9.11)(14)), we have, since $d \geq 2$, isomorphisms for $n \geq 1$,

$$
\begin{aligned}
\Gamma_n &\xrightarrow[\cong]{ad\Gamma_{n+1}} \mathrm{Aut}(\Gamma_{n+1}; \Gamma_n) \xrightarrow[\cong]{\mathrm{res}} \mathrm{Aut}(\Gamma_{n+1}'; \Gamma_n'), \\
\Gamma_0 &\xrightarrow[\cong]{ad\Gamma_\infty} \mathrm{Aut}^{\mathrm{filt}}(\Gamma_\infty) \xrightarrow[\cong]{\mathrm{res}} \mathrm{Aut}^{\mathrm{filt}}(\Gamma_\infty').
\end{aligned}
\tag{7}
$$

Moreover,

(8) Γ_∞' is a simple group, and Γ_0' normally generates $\Gamma_0'^+ = \Gamma_1'$. Hence $\hat{\Gamma}' = \{1\}$ and $\hat{\Gamma} = \Gamma/\Gamma'$.

It follows then from (6.1)(14)(7) and (10.1)(26) that

$$
C_G(\Gamma) = N_G(\Gamma') = C_G(\Gamma') = \Gamma.
\tag{9}
$$

10.6   Product groupings

Given any Nagao ray (cf. (10.1)),

(1) $\underset{\circ}{\overset{q_0+1}{\rule{0pt}{0pt}}} \rule[0.3em]{5em}{0.4pt} \underset{\circ}{\overset{q_1}{\rule{0pt}{0pt}}} \overset{1}{\rule[0.3em]{4em}{0.4pt}} \underset{\circ}{\overset{q_2}{\rule{0pt}{0pt}}} \overset{1}{\rule[0.3em]{4em}{0.4pt}} \underset{\circ}{\overset{q_s}{\rule{0pt}{0pt}}} \overset{1}{\rule[0.3em]{4em}{0.4pt}} \quad \cdots,$

we can use a "product grouping" with

$$
\begin{aligned}
\Gamma_n &= Q_1 \times \cdots \times Q_n \quad (= \{1\} \text{ if } n = 0), \\
Q_n &= \text{any group of order } q_n, \quad \text{and} \\
\Gamma_0^+ &= \text{any group of order } q_0 + 1.
\end{aligned}
\tag{2}
$$

(3)

$$\Gamma_0^+ \qquad\qquad \Gamma_1 \qquad\qquad \Gamma_2 \qquad\qquad \Gamma_3$$

$$\underset{\Gamma_0 = \{1\}}{\circ}\rule{1cm}{0.4pt}\underset{\Gamma_1}{\circ}\rule{1cm}{0.4pt}\underset{\Gamma_2}{\circ}\rule{1cm}{0.4pt}\circ\ \cdots\cdots$$

We then have

$$\Gamma = \Gamma_0^+ * \Gamma_\infty \text{ is residually finite,}$$

(4)
$$\Gamma_\infty = \bigcup_n \Gamma_n = \overset{\text{weak}}{\prod_{n \geq 1}} Q_n, \text{ and}$$

(5)
$$\mathrm{Vol}(\Gamma \backslash\backslash X) = \frac{1}{q_0 + 1} + \sum_{n \geq 1} \frac{1}{q_1 q_2 \cdots q_n}.$$

As usual $((10.1)(8))$,

(6)
$$Z_G(\Gamma) = \{1\},$$

and $((10.1)(26))$

(7)
$$N_G(\Gamma)/\Gamma \cong \mathrm{Aut}(\Gamma_0^+) \times \mathrm{Aut}^{\mathrm{filt}}(\Gamma_\infty).$$

Further, it is easy to see that

(8)
$$\mathrm{Aut}^{\mathrm{filt}}(\Gamma_\infty) = \left(\prod_{n \geq 1} \mathrm{Aut}(Q_n)\right) \times U, \text{ where}$$
$$U = \prod_{n \geq 1} Hom(Q_n, Z(Q_1 \times \cdots \times Q_{n-1})).$$

If $Q_n^{ab} = \{1\}$ (or if $Z(Q_n) = \{1\}$) for all $n \geq 1$, then $U = \{1\}$. If $q_n = q$ for all $n \geq 0$ then $X = X_{q+1}$, the $(q+1)$-*regular* tree, and we can then take all Q_n $(n \geq 1)$ to be the same group Q of order q, so that $\Gamma_n = Q^n$ for $n \geq 0$.

Appendix [BCR]

The Existence Theorem for Tree Lattices

Hyman Bass, Lisa Carbone, and Gabriel Rosenberg

Contents

0 Statement of the main results

Let X be a locally finite tree, and $G = \mathrm{Aut}(X)$. Then G is naturally a locally compact group with compact open vertex stabilizers G_x, $x \in VX$ ([BL], (3.1)). A subgroup $\Gamma \le G$ is discrete if and only if Γ_x is a finite group for some (hence for every) $x \in VX$. A discrete subgroup $\Gamma \le G$ is called an *X-lattice* if

$$(1) \qquad\qquad \mathrm{Vol}(\Gamma \backslash\backslash X) = \sum_{x \in V(\Gamma \backslash X)} \frac{1}{|\Gamma_x|}$$

is finite, and a *uniform X-lattice* if $\Gamma \backslash X$ is a finite graph.

We wish to determine conditions that will ensure that G contains X-lattices. Let μ be a (left) Haar measure on G. Let $\Gamma \le G$ be a discrete subgroup with quotient $p\ G \longrightarrow \Gamma \backslash G$. Then μ induces a measure, also denoted μ, on $\Gamma \backslash G$. We call Γ a *G-lattice* if $\mu(\Gamma \backslash G) < \infty$, and a *uniform G*-lattice if $\Gamma \backslash G$ is compact.

For $x \in VX$, we have $0 < \mu(G_x) < \infty$. When G is unimodular, $\mu(G_x)$ is constant on G-orbits, so we can define ([BL], (1.5)):

(2)
$$\mu(G \backslash\backslash X) = \sum_{x \in V(G \backslash X)} \frac{1}{\mu(G_x)}.$$

0.1   Theorem ([BL], (1.6)). *For a discrete subgroup $\Gamma \leq G = \mathrm{Aut}(X)$, the following conditions are equivalent:*

(a) *Γ is an X-lattice, that is, $\mathrm{Vol}(\Gamma \backslash\backslash X) < \infty$.*

(b) *Γ is a G-lattice (hence G is unimodular), and $\mu(G \backslash\backslash X) < \infty$.*

In this case,
$$\mathrm{Vol}(\Gamma \backslash\backslash X) = \mu(\Gamma \backslash G) \cdot \mu(G \backslash\backslash X).$$

The main result proved here is the following theorem, originally conjectured in an earlier version of [BL]:

0.2   Lattice Existence Theorem. *Let X be a locally finite tree, let $G = \mathrm{Aut}(X)$, and let μ be a (left) Haar measure on G. The following conditions are equivalent:*

(a) *G contains an X-lattice Γ.*

(b) (U)   *G is unimodular, and*
     (FV)   *$\mu(G \backslash\backslash X) < \infty$.*

0.3   Remarks

(1) The implication (a) $\Longrightarrow$ (b) follows from Theorem (0.1).

(2) When (FV) is replaced by the stronger condition

(F) $G \backslash X$ is finite,

then we have the:

0.4   Uniform Existence Theorem ([BK], (4.10)). *Let X be a locally finite tree and let $G = \mathrm{Aut}(X)$. The following conditions are equivalent:*

(a) *G contains a uniform X-lattice Γ, which is also a uniform G-lattice.*

(b) *G contains a uniform X-lattice Φ such that $\Phi \backslash X = G \backslash X$.*

(c) (U)   *G is unimodular, and*
     (F)   *$G \backslash X$ is finite.*

(d) *X is the universal cover of a finite connected graph.*

Under these conditions, X is called a uniform tree.

In light of (0.4), to prove the Lattice Existence Theorem (0.2), we are reduced to proving:

0.5   Theorem. *Let X be a locally finite tree, let $G = \text{Aut}(X)$, and let μ be a (left) Haar measure on G. Assume that*

(U) G *is unimodular,*

(FV) $\mu(G\backslash\backslash X) < \infty,$

and

(INF) $G\backslash X$ *is infinite.*

Then G contains a (necessarily non-uniform) X-lattice Γ.

This theorem will be deduced from the following result about "edge-indexed graphs." Here we follow the notations and terminology of [BL], Ch. 2, and we defer explanation until Section 2. Strictly speaking, this application presumes that G acts on X without inversions. When this is not so we can either replace X by its barycentric subdivision (cf. p. 198) or replace G by a subgroup of index ≤ 2, (cf. (5.1)(6)).

0.6   The Bounding Denominators Theorem. *Let (A, i) be an edge-indexed graph, and let $a_0 \in VA$. Assume*

$(U)_{(A,i)}$ (A, i) *is unimodular.*

Then there is a canonical covering

$$p \ (B, j) \longrightarrow (A, i)$$

of edge-indexed graphs and a $b_0 \in p^{-1}(a_0)$ with the following properties:

$(U)_{(B,j)}$ (B, j) *is unimodular.*

(FF)   *p has finite fibers. Hence B is infinite if and only if A is infinite.*

(V)
$$\text{Vol}_{b_0}(B, j) = \text{Vol}_{a_0}(A, i).$$
Hence $\text{Vol}(B, j) < \infty$ *if and only if* $\text{Vol}(A, i) < \infty.$

$(BD)_{(B,j)}$ (B, j) *has bounded denominators.*

The utility of this result is indicated by the following.

0.7   Theorem ([BK], (2.4)). *Let (B, j) be an edge-indexed graph. Then (B, j) admits a finite faithful grouping $\mathbb{B} = (B, \mathcal{B})$ if and only if (B, j) satisfies*

$(U)_{(B,j)}$ *(B, j) is unimodular*

and

$(BD)_{(B,j)}$ *(B, j) has bounded denominators.*

With the notations of (0.6) and (0.7), if $b_0 \in VB$ and $a_0 = p(b_0) \in VA$, then we put

$$(\widetilde{A, i, a_0}) = X = (\widetilde{B, j, b_0})$$

so that

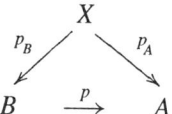

commutes.

Let $G = Aut(X)$ and

$$G_{(B,j)} = \{g \in G \mid p_B \circ g = p_B\} \le G_{(A,i)} = \{g \in G \mid p_A \circ g = p_A\}.$$

Let $\Gamma = \pi_1(\mathbb{B}, b_0) \le G_{(B,j)}$.

0.8   Theorem. *Assuming*

$(U)_{(A,i)}$ *(A, i) is unimodular*

and

$(FV)_{(A,i)}$ $\mathrm{Vol}(A, i) < \infty,$

then Γ is an X-lattice, $\Gamma \le G_{(A,i)}$, and Γ is a uniform $G_{(A,i)}$-lattice. In fact, if $x_0 \in p_B^{-1}(b_0)$, then

$$\mathrm{Vol}(\Gamma \backslash\!\backslash X) = \frac{1}{|\Gamma_{x_0}|} \mathrm{Vol}_{a_0}(A, i).$$

0.9   Corollary. *Let X be a locally finite tree, $G = \mathrm{Aut}(X)$, $H \le G$ a subgroup acting without inversions, p_H $X \longrightarrow A = H \backslash X$, and $(A, i) = I(H \backslash\!\backslash X)$. Assume that $H = G_{(A,i)}$. Then the following conditions are equivalent:*

  (a) *There is an X-lattice $\Gamma \le H$.*

  (a) $(U)_H$ *H is unimodular, and*
        $(FV)_H$ $\mu(H \backslash\!\backslash X) < \infty.$

(a) (U)$_H$ *H is unimodular, and*

(FV)$_H$ $\mu(H\backslash\backslash X) < \infty$.

Under these conditions, we can choose Γ to be a uniform H-lattice.

Proof. By Remark ((0.3), (1)), we need only verify the implication (b) $\Longrightarrow$ (a). We verify that the hypotheses of Theorem (0.8) are satisfied. We have (U)$_H$ $\Longleftrightarrow$ (U)$_{(A,i)}$ by ([BK], (3.7)), and we have (FV)$_H$ $\Longleftrightarrow$ Vol$(A, i) < \infty$ by ([BL], (3.6)(2)). Taking $H = G$ gives (0.2) and (0.5). $\qquad\qquad\qquad\square$

1   Related existence results and questions

Under the assumptions that G is unimodular, and $\mu(G\backslash\backslash X) < \infty$, Theorem (0.5) gives existence of an X-lattice Γ that is, of course, non-uniform when $G\backslash X$ is infinite. However, the Γ we produce in Theorem (0.5) is a *uniform G-lattice*. This naturally raises the following:

1.1   Question. *Let X be a locally finite tree that admits a non-uniform X-lattice. Does X admit one that is also a non-uniform G-lattice?*

Question (1.1) has a positive answer if X is a uniform tree ([C1], [C2]). Hence it remains to answer Question (1.1) in the case that $G = \text{Aut}(X)$ is unimodular. $\mu(G\backslash\backslash X) < \infty$ and $G\backslash X$ is infinite. We address this case in [CR], assuming that X has more than one end, and in [CC] when X has a unique end.

Following ([BL], (3.5)), we call X *rigid* if G is discrete, and we call X *minimal* if G acts minimally on X, that is, there is no proper G-invariant subtree. If X is uniform, then there is always a unique minimal G-invariant subtree $X_0 \subseteq X$ ([BL] (5.7), (5.11), (9.7)). We call X *virtually rigid* if X_0 is rigid (cf. ([BL], (3.6)). All X-lattices on a virtually rigid tree must be uniform ([BL], (3.6)). Conversely:

1.2   Non-Uniform Lattices on Uniform Trees ([C1], [C2]). *If X is uniform and not virtually rigid, then G contains a non-uniform X-lattice Γ, which is also (necessarily) a non-uniform G-lattice.*

We observe that the assumptions of Theorems (1.2) and (0.5) are mutually exclusive. In fact, under the conditions of Theorem (0.5), either X has a unique minimal G-invariant subtree X_0, and X_0 is not rigid, or else X is parabolic and has no rigid G-invariant subtrees.

2   Basics on edge-indexed graphs

Let (A, i) be an edge-indexed graph, in the sense of [BL], (2.5). For $e \in EA$, we put

(1) $$\Delta(e) = \frac{i(\bar{e})}{i(e)}.$$

For an edge path $\gamma = (e_1, \ldots, e_n)$ in A, we put $\Delta(\gamma) = \Delta(e_1) \cdots \Delta(e_n)$.

2.1   Proposition and definition ([BL], (2.6)). *Under the following equivalent conditions, we say that (A, i) is unimodular:*

(a) $\Delta(\gamma) = 1$ *for all closed paths γ in A.*

(b) *If γ is a path in A from a to b, then $\Delta(\gamma)$ depends only on (a, b). (We then write $\Delta(\gamma) = \frac{\Delta b}{\Delta a}$.)*

(c) *There is a function $N \; VA \longrightarrow \mathbb{R}^{\times}$ such that*

(2) $$\Delta(e) = \frac{N(\partial_1 e)}{N(\partial_0 e)} \quad \text{for all } e \in EA.$$

(Such an N is then unique up to a constant factor).

2.2   Volumes, denominators and coverings

Now assume that

$(U)_{(A,i)}$ (A, i) is unimodular.

Pick a base point $a_0 \in VA$, and define, for $a \in VA$,

(1) $$N_{a_0}(a) = \frac{\Delta a}{\Delta a_0} \; (= \Delta(\gamma) \text{ for any path } \gamma \text{ from } a_0 \text{ to } a) \in \mathbb{Q}_{>0}.$$

Then N_{a_0} is the function N as in (2.1)(c) such that $N(a_0) = 1$.
     For $e \in EA$, put

(2) $$N_{a_0}(e) = \frac{N_{a_0}(\partial_0(e))}{i(e)}.$$

Then the condition (2) in (2.1) is equivalent to

(3) $$N_{a_0}(e) = N_{a_0}(\overline{e}).$$

Following [BL], (2.6), we say that

$(BD)_{(A,i)}$ (A, i) has bounded denominators.

if $\{N_{a_0}(e) \mid e \in EA\}$ has bounded denominators, that is, if for some integer $D > 0$, $D \cdot N_{a_0}$ takes only integer values on edges. Since

(4) $$N_{a_1} = \frac{\Delta a_0}{\Delta a_1} N_{a_0},$$

this condition is independant of $a_0 \in VA$.

Following [BL], (2.6), we define the *volume* of an edge-indexed graph (A, i) at a base point $a_0 \in VA$:

$$(5) \qquad \mathrm{Vol}_{a_0}(A, i) = \sum_{a \in VA} \frac{1}{N_{a_0}(a)} = \sum_{a \in VA} \left(\frac{\Delta a_0}{\Delta a} \right).$$

From (4), we have

$$\mathrm{Vol}_{a_1}(A, i) = \frac{\Delta a_1}{\Delta a_0} \mathrm{Vol}_{a_0}(A, i),$$

so the condition

$$(FV)_{(A,i)} \qquad\qquad\qquad \mathrm{Vol}(A, i) < \infty,$$

defined by $\mathrm{Vol}_{a_0}(A, i) < \infty$, is independant of the choice of a_0.

Moreover, (A, i, a_0) admits a covering tree $X = \widetilde{(A, i, a_0)}$ with a projection

$$p_{(A,i)} \ X \longrightarrow A$$

([BL], (2.5)). We put $G = \mathrm{Aut}(X)$ and $G_{(A,i)} = \{g \in G \mid p_{(A,i)} \circ g = p_{(A,i)}\}$.

Finally, we explain the notion of a *covering* of edge-indexed graphs ([BL], (2.5)),

$$p \ (B, j) \longrightarrow (A, i).$$

Here $p \ B \longrightarrow A$ is a graph morphism such that for all $e \in EA$, $\partial_0(e) = a$, and $b \in p^{-1}(a)$, we have

$$(6) \qquad\qquad\qquad i(e) = \sum_{f \in p_{(b)}^{-1}(e)} j(f),$$

where $p_{(b)} \ E_0^B(b) \longrightarrow E_0^A(a)$ is the local map on stars $E_0^B(b)$ and $E_0^A(a)$ of vertices $b \in VB$ and $a \in VA$ (cf. [BL], (2.5)). If $b \in VB$, $p(b) = a \in VA$, then we can identify

$$\widetilde{(A, i, a)} = X = \widetilde{(B, j, b)}$$

so that the diagram of natural projections

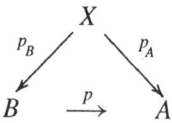

commutes. Hence $G_{(B,j)} \leq G_{(A,i)}$.

3   The Bounding Denominators Theorem: Examples and Applications

Here we prove the Bounding Denominators Theorem (0.6). In (3.1), we first show how to make the indices on edges relatively prime with the "caging covering." In (3.2), we prepare some arithmetic notation. Then in (3.3), we prove the Bounding Denominators Theorem. In (3.4), we illustrate some examples. Finally, in (3.5)–(3.7), we give the main application.

3.1   Caging to make edge-indices relatively prime

Let (A, i) be an edge-indexed graph. For $e \in EA$, put

(1)
$$\delta(e) = gcd(i(e), i(\overline{e})), \quad \text{and}$$
$$i(e) = \delta(e)i_0(e) \quad \text{so that} \quad gcd(i_0(e), i_0(\overline{e})) = 1.$$

We construct the *caging covering*:

(2)
$$(/\delta) \; (A', i') \longrightarrow (A, i)$$

as follows. (The notation "$(/\delta)$" is meant to suggest division by δ.)

(3) $\qquad\qquad VA' = VA \quad \text{and} \quad (/\delta) \mid_{VA'} = Id,$

(4) $\qquad\qquad EA' = \{e(k) \mid e \in EA, \; 1 \leq k \leq \delta(e)\},$

$$\partial_0 e(k) = \partial_0 e, \qquad \partial_1 e(k) = \partial_1 e, \qquad \overline{e(k)} = \overline{e}(k),$$
$$i'(e(k)) = i_0(e), \qquad (/\delta)(e(k)) = e.$$

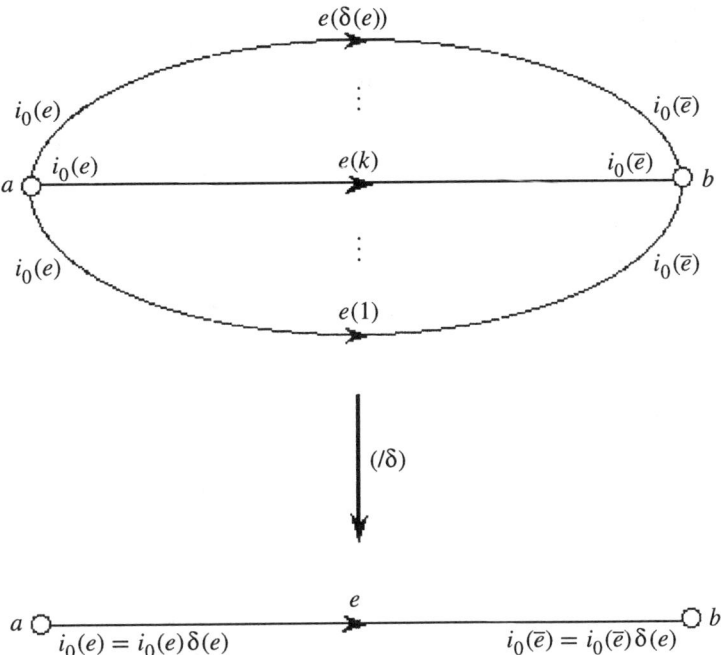

In (A', i') we have

$$\text{(5)} \qquad \Delta'(e(k)) = \frac{i'(\overline{e(k)})}{i'(e(k))} = \frac{i_0(\overline{e})}{i_0(e)} = \frac{i(\overline{e})}{i(e)} = \Delta(e).$$

Thus, if γ is an edge path in A', and $\gamma = (/\delta)(\gamma')$, then $\Delta'(\gamma') = \Delta(\gamma)$. It follows that

(6) $\qquad (A', i')$ is unimodular if and only if (A, i) is unimodular.

Assume that this holds. Then further, for $a_0, a \in VA$,

$$\text{(7)} \qquad \frac{\Delta a}{\Delta a_0} \text{ is the same, calculated in } (A, i) \text{ or } (A', i').$$

It follows that

$$\text{(8)} \qquad \text{Vol}_{a_0}(A', i') = \text{Vol}_{a_0}(A, i).$$

Finally, with regard to bounding denominators, consider $e \in EA$ with $\partial_0 e = a$. Then

$$\text{(9)} \qquad \left(\frac{\Delta a}{\Delta a_0}\right) / i(e(k)) = \left(\frac{\Delta a}{\Delta a_0}\right) / i_0(e) = \left(\left(\frac{\Delta a}{\Delta a_0}\right) / i(e)\right) \cdot \delta(e).$$

Thus (A', i') has the effect of removing $\delta(e)$ from the denominator. It follows that while bounded denominators for (A, i) implies the same for (A', i'), the converse need not be true if the $\delta(e)'s$ are not bounded.

3.2   Some arithmetic index notation

Let (A, i) be a unimodular edge-indexed graph. Fix a base point $a_0 \in VA$. For an edge $e \in EA$ recall that $\Delta(e) = i(\overline{e})/i(e)$. For a path $\gamma = (e_1, \dots, e_n)$ from a_0 to a, $\Delta(\gamma) = \Delta(e_1) \cdots \Delta(e_n)$ depends only on a_0 and a, not on γ. We denote this

(1)
$$\frac{\Delta a}{\Delta a_0} = \frac{c(a)}{d(a)} \quad \text{in reduced form.}$$

Note that $c(a_0) = d(a_0) = 1$. If e is an edge from a to a', then (γ, e) is a path from a_0 to a', so we see that

(2)
$$\text{If } \partial_0 e = a, \ \partial_1 e = a', \text{ then}$$
$$\Delta(e) = \left(\frac{\Delta a'}{\Delta a_0}\right) \Big/ \left(\frac{\Delta a}{\Delta a_0}\right) \quad = \frac{c(a')d(a)}{c(a)d(a')}.$$

Write

(3)
$$\begin{array}{ll} c(a, a') = gcd(c(a), c(a')), & c(a) = c_{a'}(a)c(a, a'), \\ d(a, a') = gcd(d(a), d(a')), & d(a) = d_{a'}(a)d(a, a'). \end{array}$$

Then we see from (1) and (2) that

(4)
$$\Delta(e) = \frac{c_a(a')d_{a'}(a)}{c_{a'}(a)d_a(a')} \quad \text{in reduced form}$$

so that in the notation of (3.1), where $\Delta(e) = i_0(\overline{e})/i_0(e)$ in reduced form, we conclude that

(5)
$$\begin{array}{l} i_0(\overline{e}) = c_a(a')d_{a'}(a), \\ i_0(e) = c_{a'}(a)d_a(a'). \end{array}$$

Further, in connection with bounding denominators, we have

(6)
$$\left(\frac{\Delta a}{\Delta a_0}\right) \Big/ i(e) \overset{(1)}{=} \frac{c(a)}{d(a)i(e)} \overset{(3),(3.1)(1)}{=} \frac{c_{a'}(a)c(a, a')}{d(a)i_0(e)\delta(e)}$$

$$\overset{(5)}{=} \frac{c(a, a')}{d(a)d_a(a')\delta(e)}$$

$$\overset{(3)}{=} \left(\frac{gcd(c(a), c(a'))}{lcm(d(a), d(a'))}\right) \Big/ \delta(e).$$

3.3   Construction of the denominator clearing covering

As in (3.2), let (A, i) be a unimodular edge-indexed graph with a base point $a_0 \in VA$. We use the notations of (3.1) and (3.2). Our objective now is to construct a covering,

(1) $p^*\ (A^*, i^*) \longrightarrow (A, i)$

as follows:

(2) $VA^* = \{b_a(k) \mid a \in VA,\ 0 \le k < d(a)\},\quad p^*(b_a(k)) = a,$

(3) $EA^* = \{f_e(k, r') \mid e \in EA,$

$\qquad\qquad$ say $\partial_0 e = a,\ \partial_1 e = a',\ 0 \le k < d(a),\ 0 \le r' < d_a(a')\}.$

Let $e \in EA,\ \partial_0 e = a,\ \partial_1 e = a'$. We set

(4) $p^*(f_e(k, r')) = e,$

(5) $\partial_0 f_e(k, r') = b_a(k),$

(6) $i^*(f_e(k, r')) = c_{a'}(a)\delta(e).$

Write

(7) $k = \langle q, r \rangle_{d_{a'}(a)} = q \cdot d_{a'}(a) + r,\quad 0 \le r < d_{a'}(a)$

(Euclidean division), so that we can write $f_e(k, r') = f_e(\langle q, r \rangle_{d_{a'}(a)}, r')$.

$\quad$ Putting

(8) $k' = \langle q, r' \rangle_{d_a(a')} = q \cdot d_a(a') + r',$

this gives Euclidean division of k' by $d_a(a')$, and we define

(9) $\overline{f_e(k, r')} = f_{\bar e}(k', r)$

and

(10) $\partial_1 f_e(k, r') = b_{a'}(k').$

$\quad$ To check the graph axioms, we note that $(k, r') \leftrightarrow (k', r)$ define inverse bijections as follows. For an integer $d > 0$, put $d = \{0, 1, 2, \ldots, d - 1\}$. Then

$$(k, r') \in d(a) \times d_a(a')$$

and

$$(k', r) \in d(a') \times d_{a'}(a).$$

Euclidean division gives bijections

$$d(a) \xrightarrow{\cong} d(a, a') \times d_{a'}(a),$$
$$k = \langle q, r \rangle_{d_{a'}(a)} \longrightarrow (q, r),$$

and similarly for $d(a')$. Thus the bijections above correspond to

$$(d(a, a') \times d_{a'}(a)) \times d_a(a') \leftrightarrow (d(a, a') \times d_a(a')) \times d_{a'}(a),$$
$$((q, r), r') \leftrightarrow ((q, r'), r).$$

From this, the graph axioms on A^* (for $f \in EA^*$, $\partial_j \overline{f} = \partial_{1-j} f$ ($j = 0, 1$), $\overline{f} \neq f = \overline{\overline{f}}$) are easily verified. For the local map

$$p^*_{(b_a(k))} \ E_0(b_a(k)) \longrightarrow E_0(a)$$

and $e \in E_0(a)$, we have

$$\sum_{f \in p^{*-1}_{(b_a(k))}(e)} i^*(f) \overset{(3),(4)}{=} \sum_{0 \leq r' < d_a(a')} i^*(f_e(k, r'))$$

$$\overset{(6)}{=} \sum_{0 \leq r' < d_a(a')} c_{a'}(a) \cdot \delta(e)$$

$$= c_{a'}(a) d_a(a') \delta(e)$$

$$\overset{(3.2)(5)}{=} i_0(e) \delta(e)$$

$$\overset{(3.1)(1)}{=} i(e),$$

and so

(11) p^* is a covering.

Let $e \in EA$, $\partial_0 e = a$, $\partial_1 e = a'$. Then we have, in (A^*, i^*),

(12) $\Delta^*(f_e(k, r')) \overset{(9)}{=} \dfrac{i^*(f_{\overline{e}}(k', r))}{i^*(f_e(k, r'))}$

$$\overset{(6)}{=} \dfrac{c_a(a') \delta(e)}{c_{a'}(a) \delta(e)}$$

$$\overset{(3.2)(3)}{=} \dfrac{c(a')}{c(a)}.$$

Hence if γ^* is a path in A^* from b to b_1, then $p^*\gamma^*$ is a path in A from $p(b) = a$ to $p^*(b_1) = a_1$, and $\Delta^*(\gamma^*) = c(a_1)/c(a)$. This implies that

(13) (A^*, i^*) is unimodular

and that, if we put $b_0 = b_{a_0}(0)$,

(14)
$$\frac{\Delta(b_a(k))}{\Delta b_0} = c(a) \quad (\in \mathbb{Z}).$$

Further,

(15)
$$\mathrm{Vol}_{b_0}(A^*, i^*) \overset{(14)}{=} \sum_{a \in VA} \sum_{0 \le k < d(a)} \frac{1}{c(a)}$$

$$= \sum_{a \in VA} \frac{d(a)}{c(a)}$$

$$\overset{(3.2)(1)}{=} \sum_{a \in VA} \frac{\Delta a_0}{\Delta a}$$

$$= \mathrm{Vol}_{a_0}(A, i).$$

Finally, for $f_e(k, r') \in EA^*$, $\partial_0 e = a$, $\partial_1 e = a'$, we have

(16)
$$\left(\frac{\Delta b_a(k)}{\Delta b_0} \right) / i^*(f_e(k, r')) \overset{(14),(6)}{=} \frac{c(a)}{c_{a'}(a)\delta(e)}$$

$$\overset{(3.2)(3)}{=} \frac{c(a, a')}{\delta(e)}.$$

It follows that

(17)   If $\{\delta(e) \mid e \in EA\}$ is bounded (for example, if $\delta = 1$), then (A^*, i^*) has bounded denominators.

Note further that in (A^*, i^*),

(18)
$$\delta^*(f_e(k, r')) \overset{(9)}{=} \gcd(i^*(f_e(k, r')), i^*(f_{\bar{e}}(k', r)))$$

$$\overset{(6)}{=} \gcd(c_{a'}(a)\delta(e), c_a(a')\delta(\bar{e}))$$

$$\overset{(3.2)(3)}{=} \delta(e).$$

Now we can use the "caging covering" of (3.1) to get rid of the denominators $\delta(e)$ in (16) and so achieve bounded denominators—*in fact, clear denominators*—for (A^*, i^*). This can be applied either at the end, to (A^*, i^*), or else at the beginning, to (A, i). Note from (3.2) that this caging covering preserves unimodularity $((3.2)(6))$ and volume $((3.2)(8))$.

Let $(/\delta)$ $(B, j) \longrightarrow (A^*, i^*)$ be the caging covering (3.2), and let

(19)
$$p = p^* \circ (/\delta) \ (B, j) \longrightarrow (A, i).$$

Then by (11) and (3.2),

(FF) p is a covering, with finite fibers,

and by (13) and (3.1)(6),

(U) (B, j) is unimodular.

Moreover, by (15) and (3.1)(8),

(V) $\mathrm{Vol}_{b_0}(B, j) = \mathrm{Vol}_{a_0}(A, i)$.

Finally, from (16), (18), and (3.1), it follows that for $f \in EB$ above $f_e(k, r') \in EA^*$, we have, in (B, j),

$$
(20) \quad \left(\frac{\Delta b_a(k)}{\Delta b_0}\right) / j(f) \;=\; \left(\left(\frac{\Delta^* b_a(k)}{\Delta^* b_0}\right) / i^*(f_e(k, r'))\right) \cdot \delta(f_e(k, r'))
$$

$$
\overset{(16),(18)}{=} \left(\frac{c(a, a')}{\delta(e)}\right) \cdot \delta(e)
$$

$$
= \quad c(a, a') \in \mathbb{Z}.
$$

Thus

(BD) (B, j) has bounded (in fact, no) denominators.

We have verified (FF), (U), (V), and (BD) for (B, j), and this completes the proof of the Bounding Denominators Theorem (0.6).

3.4   Examples

We shall illustrate the construction of (3.3) with the indexed triangle (A, i) given in Figures 0 and 1 and defined as follows:

(1) $VA = \{0, 1, 2\}, \qquad EA = \{(s, t) \mid 0 \le s, t \le 2, \; s \ne t\}$
                with $\partial_0(s, t) = s$ and $\overline{(s, t)} = (t, s)$.

(2) $i(0, 1) = 2 = i(2, 1), \qquad i(1, 0) = 1 = i(2, 0), \qquad i(0, 2) = 3 = i(1, 2)$.

We have $\Delta(0, 1) = 1/2$, $\Delta(1, 2) = 2/3$, and $\Delta(2, 0) = 3/1$, and so (A, i) is unimodular. The edge indices in (A, i) are relatively prime, so the covering

(3) $p \; (B, j) \longrightarrow (A, i)$

constructed in (3.3) coincides with p^*: $(A^*, i^*) \longrightarrow (A, i)$. This construction depends on the choice of a base vertex a_0 in A. Figures 0 and 1 illustrate this covering relative to the choice of $a_0 = 0$ and 1, respectively.

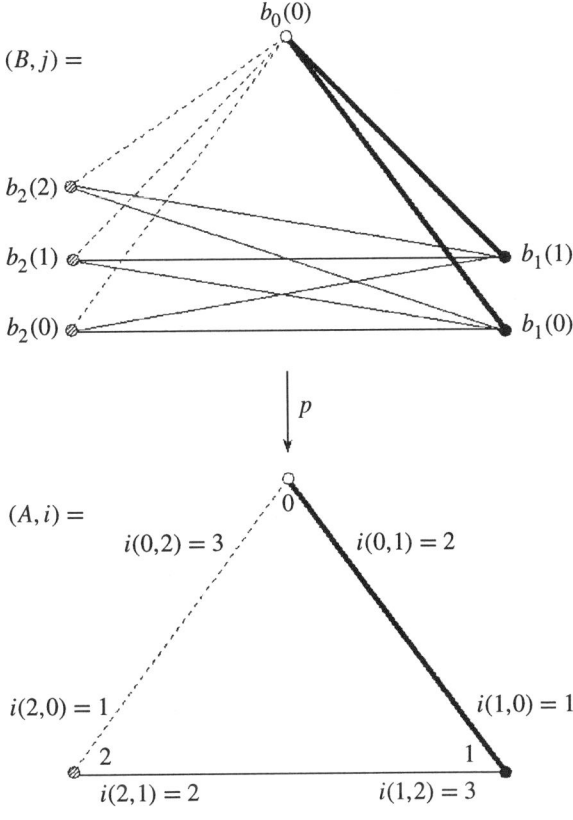

$(B, j) =$

$(A, i) =$

Figure 0.

Choosing 0 as our base vertex, we have, in the notation of (3.2)(1),

(4) $c(0) = c(1) = c(2) = 1 = d(0)$, $d(1) = 2$,   and $d(2) = 3$.

Thus, in this case, B will have $v + 1$ vertices above vertex v ($v = 0, 1, 2$) (see Figure 0), and (B, j) will be unramified. (An edge e is said to be *unramified* if $i(e) = 1$.)

Figure 1, with 1 as base vertex, shows that B then has only one vertex above each of 0 and 1, but three vertices above 2. Moreover, (B, j) had index 2 on the edges above (0,1) and (2,1) but is otherwise unramified.

3.5   Grouping (B, j)

Let (B, j) be a unimodular edge-indexed graph with covering tree X,

(1) $p_B \ X \longrightarrow B$,

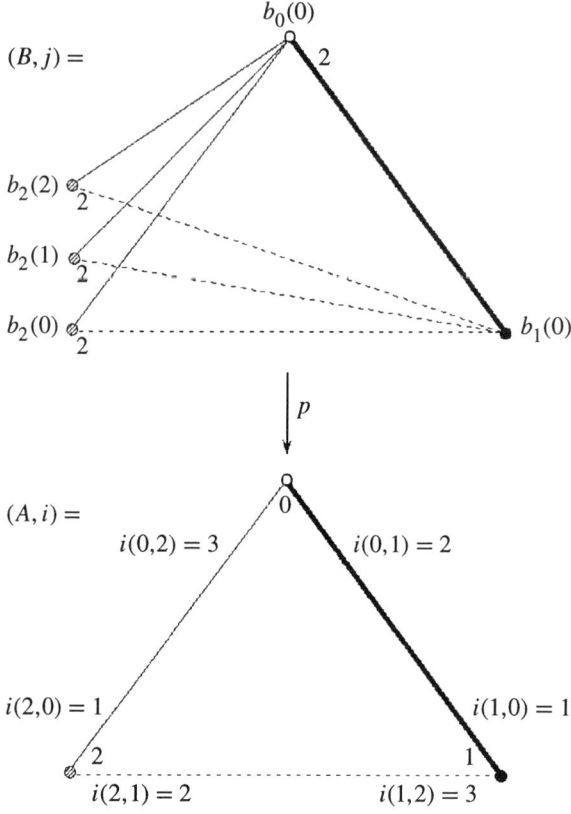

Figure 1.

and $G = \mathrm{Aut}(X)$. Choose a base point x_0 in X and put $b_0 = p_B(x_0)$. Consider the order function

(2) $N(b) = \Delta(b)/\Delta(b_0)$

on VB (so that $N(b_0) = 1$) and, for f in EB with $\partial_0(f) = b$,

(3) $N(f) = N(b)/j(f)$.

Assume

(4) $N(f) \in \mathbb{Z}$ for all f in EB.

It follows that $N(b)$ is likewise an integer for all vertices b of B. Now, in the terminology of [BL], (2.5)(2), we can make a

(5)   *finite grouping* $\mathbb{B} = (B, \mathcal{B})$ *of* (B, j)

as follows. Choose $\mathcal{B}_b$ ($b \in VB$) and $\mathcal{B}_f$ ($f \in EB$) to be groups of orders $N(b)$ and $N(f)$, respectively. If b is the initial vertex of f it remains to construct an injection of $\mathcal{B}_f$ into $\mathcal{B}_b$. What we know is that $N(f)$ divides $N(b)$. Therefore, this problem is solved provided we choose our groups from a family $\mathcal{G}$ of groups such that, whenever H and H' are groups in $\mathcal{G}$ and $|H|$ divides $|H'|$, then H admits an embedding into H'.

For this purpose, we could take for $\mathcal{G}$ either the class $\mathcal{C}$ of finite cyclic groups or else the class $\mathcal{E}$ of "elementary abelian" groups, that is, finite abelian groups in which all elements have square free order. Such a grouping $\mathbb{B}$ is automatically faithful, since $\mathcal{B}_b$ is the trivial group (cf. [B]).

3.6   Application

Let $\Gamma = \pi_1(\mathbb{B}, b_0)$. Then Γ is a discrete subgroup of G with quotient graph $B = \Gamma \backslash X$, stabilizers in $\mathcal{G}$, and $\Gamma_{x_0} = 1$.

Let (A, i) be a unimodular edge-indexed graph with universal covering tree X, $G = \text{Aut}(X)$, $H = G_{(A,i)}$, and projection p_A $X \longrightarrow A = H\backslash X$. Let x_0 be a base point in X and $a_0 = p_A(x_0)$. Let

$$p \ (B, j) \longrightarrow (A, i)$$

be the covering constructed in (3.3). Then we can identify X with the covering tree of (B, j) as well, so that $p_A = p \circ p_B$. Put $b_0 = p_B(x_0)$, which lies over a_0. Let N be as in (2) above. Then assumption (4) above holds by (3.3)(20), so we can construct the $\mathcal{G}$-grouping $\mathbb{B}$ of (B, j) as above and the discrete group Γ of $G_{(B,j)} \leq G_{(A,i)}$. Since p $B = \Gamma \backslash X \longrightarrow A = H\backslash X$ has finite fibers, it follows from [BL], (1.5)(8) that Γ is a uniform H-lattice. Thus, in view of [BL], (3.2)(3)–(4), we have proved:

3.7   Theorem. *Let (A, i) be a unimodular edge-indexed graph with universal covering tree X, $G = \text{Aut}(X)$, and let x_0 be a base point in X. Let $\mathcal{G}$ be either the class of finite cyclic groups or the class of elementary finite abelian groups. Then there is a discrete subgroup Γ of $H = G_{(A,i)}$ such that $\Gamma \backslash X \longrightarrow A = H\backslash X$ has finite fibers, Γ is a uniform H-lattice, all stabilizers Γ_x are in $\mathcal{G}$, and $\Gamma_{x_0} = 1$.*

(a) *Γ is an X-lattice $\iff$ $\text{Vol}(A, i) < \infty$ and $\mu(H\backslash\backslash X) < \infty$.*

(b) *Γ is a uniform X-lattice $\iff$ A is finite.*

Acknowledgments

This paper was completed during the Asymptotic Group Theory Program at the Institute for Advanced Studies at the Hebrew University of Jerusalem, whose warm hospitality is gratefully acknowledged.

The first and third authors were supported in part by NSF grant #DMS-9700634, and the first author was also supported by a US–Israel Binational Science Foundation (BSF) grant. The second author was supported in part by NSF grant #DMS-9800604.

References

[B] H. Bass, Covering theory for graphs of groups, *J. Pure Appl. Algebra*, **89** (1993), 3–47.

[BK] H. Bass and R. Kulkarni, Uniform tree lattices, *J. Amer. Math. Soc.*, **3** (1990), 843–902.

[BL] H. Bass and A. Lubotzky, *Tree Lattices*, Progress in Mathematics 176, Birkhäuser, Boston, 2000 (this volume).

[C1] L. Carbone, Non-uniform lattices on uniform trees, *Mem. Amer. Math. Soc.*, 2000, to appear.

[C2] L. Carbone, Non-minimal tree actions and the existence of non-uniform tree lattices, 2000, in preparation.

[CR] L. Carbone and G. Rosenberg, Lattices on non-uniform trees, 2000, in preparation.

[CC] L. Carbone and D. Clark, Lattices on parabolic trees, 2000, preprint.

Appendix [BT]

Discreteness Criteria for Tree Automorphism Groups

Hyman Bass and Jacques Tits

Contents

Introduction

Let X be a locally finite tree (each vertex has finite degree). Then $G = \mathrm{Aut}(X)$ is naturally a locally compact group, in which the stabilizer G_x of a vertex x is open and compact (in fact profinite). When is G discrete? That is the question we treat here.

Of course G is discrete if it is finite (for example if X is finite). And, in fact, a randomly constructed locally finite tree will have no non-trivial automorphisms. At the other extreme, the most symmetric trees are the d-regular trees X_d (each vertex has degree d), where $d = 0, 1, 2, 3, \ldots$, for instance

$$X_0 = \circ, \quad X_1 = \circ\!\!\!-\!\!\!\!-\!\!\!\!-\!\!\!\!-\!\!\!\!-\!\!\!\!-\!\!\!\!-\!\!\!\!-\!\!\!\circ, \text{ and}$$

$$X_2 = -\circ\!\!\!-\!\!\!\!-\!\!\!\circ\!\!\!-\!\!\!\!-\!\!\!\circ\!\!\!-\!\!\!\!-\!\!\!\circ\!\!\!-\!\!\!\!-\!\!\!\circ\!\!\!-\!\!\!\!-\!\!\!\circ- \cdots,$$

the (bi-infinite) line. Putting $G_d = \text{Aut}(X_d)$, G_0 and G_1 are finite, and G_2 is the infinite dihedral group, which is discrete. However, for $d \geq 3$, G_d is very large. For example, for a prime p, G_{p+1} contains both of the non-archimedean Lie groups $PGL_2(\mathbb{Q}_p)$ and $PGL_2(\mathbb{F}_p((t)))$. For any vertex x of X_d $(G_d)_x$ is an infinite iterated wreath product of symmetric groups.

One's first intuition suggests perhaps that, if X is infinite, "highly symmetric," and not "virtually linear" (like X_2), then $G = \text{Aut}(X)$ should be "large," in particular not discrete.

A natural class of highly symmetric trees are the *uniform trees*. We call X uniform if $X = \tilde{A}$, the universal cover of a finite connected graph A. In this case the subgroup $\Gamma = \pi_1(A)$ of G is a free group that acts freely on X with quotient $A = \Gamma \backslash X$. Hence G (like Γ) has only finitely many orbits. If Γ is cyclic then we say that X is "virtually linear." Apart from these cases one might expect G to be large.

However this is not always so. The first examples were exhibited in [BK] (Example (4.12)2). A simpler example is

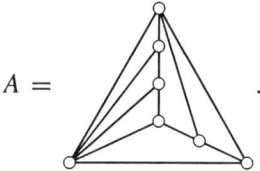

In this case we have

$$G = \Gamma = \pi_1(A),$$

a free group of rank 6. Such examples are discussed in Section 7 below.

We provide here an algorithm to decide, for a uniform tree X, when $G = \text{Aut}(X)$ is discrete, say assuming that X is presented as the universal cover $\tilde{A}$ of a finite connected graph A. In fact the method, which we now describe, answers a more general question, which has interesting applications ([BL], Chapter 8).

Our conventions and notations for graphs follow those of [BL], Chapter 2. Thus, a graph A is comprised of a set VA of vertices, a set EA of oriented edges, endpoint maps $\partial_0, \partial_1 : EA \to VA$, and orientation reversal $EA \to EA$, $e \mapsto \bar{e}$, such that $\bar{\bar{e}} = e \neq \bar{e}$ for all e in EA. For $a \in VA$, we put $E_0(a)(= E_0^A(a)) = \{e \in EA | \partial_0 e = a\}$, and $\deg(a) = |E_0(a)|$ (cardinal of $E_0(a)$). We call A locally finite if $\deg(a)$ is finite for all a.

Let X be a locally finite tree, H a group acting on X without inversion ($H \cdot e \neq H \cdot \bar{e}$ for all $e \in EX$), and

$$p : X \to A = H \backslash X$$

the quotient graph. For $x \in VX$ and $a = p(x)$, the local map $p_{(x)} : E_0^X(x) \to E_0^A(a)$ is the quotient by the local action of the stabilizer H_x on $E_0^X(x)$. Let $f \in E_0^X(x)$ and $e = p(f) \in E_0^A(a)$. Then we put

$$i(e) = |p_{(x)}^{-1}(e)| = |H_x \cdot f| = [H_x : H_f].$$

In fact this number depends only on e, and not on the choice of x and f. The resulting map

$$i : EA \to \mathbb{N} - \{0\}$$

makes (A, i) what we call an *edge-indexed graph*, which we denote by

$$(A, i) = I(H \backslash\backslash X).$$

(By an edge-indexed graph we mean simply a graph A together with a function $i : EA \to \mathbb{N} - \{0\}$. The notation $H \backslash\backslash X$ will be explained below.)

The point of this is that (A, i) suffices to determine X, as a suitably defined "universal cover,"

$$p : X = \widetilde{(A, i)} \to A.$$

The idea is that vertices of X can be taken to be non-reversing paths from a (fixed) base point in A, with the understanding that each time one leaves a vertex along an oriented edge e, there are $i(e)$ choices for this e-direction in X, so that paths are counted with these multiplicities. The universal cover of $(A, i) = \underset{2}{\overset{}{\rightleftharpoons}}$ (all undenoted indices equal 1) is illustrated below.

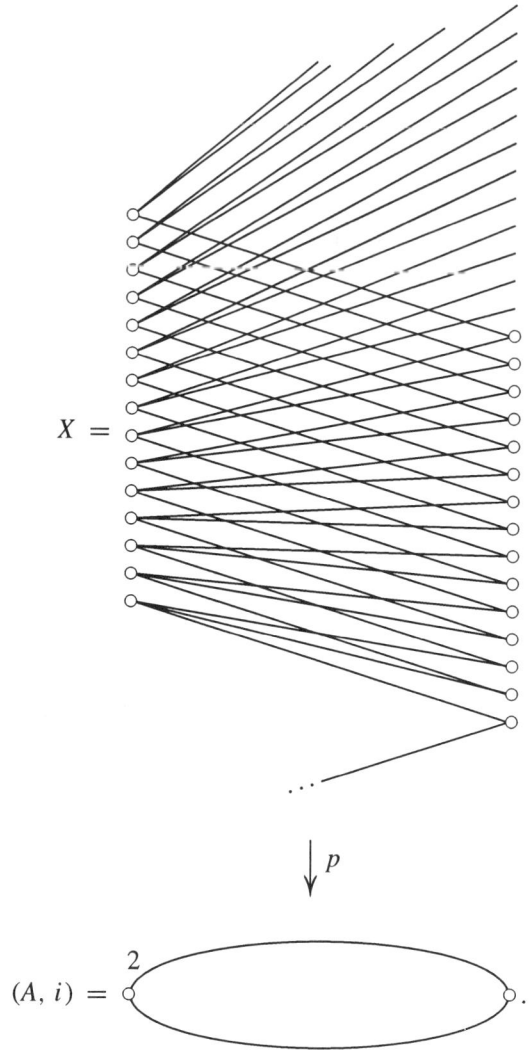

$$X =$$

$$\downarrow p$$

$$(A, i) =$$

While the edge-indexed graph $(A, i) = I(H \backslash\backslash X)$ determines the tree $X = \widetilde{(A, i)}$, it does not determine the group H, with its action on X. For this one requires a "quotient graph of groups," $H \backslash\backslash X = (A, \mathcal{A})$. Here $\mathcal{A}$ assigns groups $\mathcal{A}_a$ to each $a \in VA$ and $\mathcal{A}_e = \mathcal{A}_{\bar{e}}$ to each $e \in EA$, together with monomorphisms $\alpha_e : \mathcal{A}_e \to \mathcal{A}_{\partial_0 e}$. This graph of groups determines $I(H \backslash\backslash X)$, by the formula,

$$(*) \qquad\qquad\qquad i(e) = [\mathcal{A}_{\partial_0 e} : \alpha_e \mathcal{A}_e].$$

Any (abstract) graph of groups with this property $(*)$ is called a "grouping of (A, i)." The construction of a quotient graph of groups $H \backslash\backslash X$ depends on various choices, such as connected fundamental domains for the action of H on VX and on EX. For a vertex x or edge e of these fundamental domains one chooses for $\mathcal{A}$ the stabilizers $\mathcal{A}_{p(x)} = H_x$ and $\mathcal{A}_{p(e)} = H_e$, and then $\alpha_{p(e)}$ is the inclusion $H_e \to H_{\partial_0 e}$, followed by conjugation by a chosen element of H carrying $\partial_0 e$ into the fundamental domain for vertices. This shows that $(*)$ is consistent with our definition of $i(e)$ given above.

The main point of this theory (cf. [B]) is that any graph of groups $\mathcal{A} = (A, \mathcal{A})$ has, relative to a choice of base point a_o in VA, a universal covering tree $X = \widetilde{(A, a_0)} = (\tilde{A}, i, a_0)$, and a fundamental group $H = \pi_1(A, a_0)$ acting on X, in such a way that A can be identified with a quotient graph of groups $H \backslash\backslash X$. Conversely, starting with a group H acting (without inversion) on a tree X as above, we can naturally identify X, H and the action with the analogous data arising from a quotient graph of groups $H \backslash\backslash X$. Thus, graphs of groups furnish a compact way of encoding group actions (without inversions) on trees.

With the above notation, and $G = \mathrm{Aut}(X)$, we define the group of "deck transformations,"

$$G_{(A,i)} = \{g \in G | p \circ g = p\},$$

i.e., the groups of p-fiber preserving automorphisms of X. Our first result (Theorem (1.3)) is a discreteness criterion for $G_{(A,i)}$.

To then decide when G itself is discrete we need only apply this criterion to $I(G \backslash\backslash X) = (A_*, i_*)$. (We may have to subdivide X to arrange that G acts without inversion; see Section 5.) Then the task is to determine (A_*, i_*) directly from (A, i). This is achieved using a "degree refinement algorithm." (See Section 6.)

The theory and formalism described above are substantially similar to the ideas of [T], §5. There one considers a tree X and a function $p : VX \to I$, for some set I. In $G = \mathrm{Aut}(X)$ we then have the subgroup $G_p = \{g \in G | p \circ g = p\}$, and p is called "normal" if it is surjective and if G_p acts transitively on its fibers. For $(i, j) \in I \times I$, $i = p(x)$, we can then define the cardinal $a(i, j) = |\{e \in E_0(x) | p(\partial_1 e) = j\}|$, which is independent of the choice of x. This function a is then an edge-indexing of the "q-graph" A defined by $VA = I$ and $EA = \{(i, j) \in I \times I | (a(i, j) > 0\}$. The expression "$q$-graph" refers to the possibility of "self-inverse loops," (i, i) in A, with $a(i, i) > 0$. (In fact these are excluded from A in [T] (5.3).) The map p then extends to a morphism $p : X \to A$, with $p(e) = (p(\partial_0 e), p(\partial_1 e))$ for $e \in EX$, and p may be identified with the quotient projection mod G_p. In the notation above we then have $X = \widetilde{(A, a)}$ and $G_p = G_{(A,a)}$.

Conversely, [T] (5.3) can be interpreted as saying that, given any connected edge-indexed q-graph (A, a) (with $EA \subset VA \times VA$), there is an "a-covering tree" $p : X \to A$, unique up to isomorphism over A, such that $p : VX \to VA$ is a normal function that, as above, gives rise to (A, a).

In this translation, the notion ([T], (5.6)) of a "dominant map" corresponds to that here of a "covering of edge-indexed (q-) graphs."

The emphasis in [T] was on describing a large variety of trees with abundant symmetry and, in particular, some big locally compact groups (including many simple ones) as a kind of wild outgrowth of p-adic Lie groups of rank 1.

The graph-of-groups approach adopted here is well suited to treating finitely generated and discrete groups, and can be viewed as the counterpart here to the theory of arithmetic groups and lattices.

1 The discreteness criterion

1.1 Ramified and separating edges

Let (A, i) be an edge-indexed graph (always assumed connected). Call $e \in EA$ *ramified* if $i(e) > 1$, and *unramified* if $i(e) = 1$. Call (A, i) *unramified* if all $e \in EA$ are unramified. Call $e \in EA$ *separating* if removal of e (and $\bar{e}$) disconnects A. For $j = 0, 1$, we then denote by $A_j(e)$ the connected component of $A - \{e, \bar{e}\}$ containing the vertex $\partial_j e$.

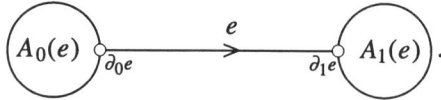

1.2 Definition of (DR)

Call (A, i) *discretely ramified* (DR) if, for all $e \in EA$,

$$[i(e) > 1] \Rightarrow \begin{cases} i(e) = 2; \\ e \quad \text{is a separating edge; and} \\ (A_0(e), i) \quad \text{is an unramified tree.} \end{cases}$$

Theorem 1.3. *Let (A, i) be a (connected) edge-indexed graph, $X = \widetilde{(A, i)}$, and $H = G_{(A, i)} \subseteq G = Aut(X)$. The following conditions are equivalent.*

(a) *H is discrete.*

(b) *Either*

    (F) *H is finite,*

or

    (DR) *(A, i) is discretely ramified.*

Further, (F) is equivalent to:

(F′) *For some $a \in VA$, (A, i, a) is a dominant-rooted edge-indexed tree with only finitely many ramified edges. (Terminology explained in Section 2.)*

Finally, (DR) implies that

$$H \cong \left(\underset{\substack{e \in EA \\ i(e)=2}}{*} \langle s_e \rangle \right) * \pi_1(A).$$

where $\pi_1(A)$ is a free group and each s_e has order two. Moreover, if H' is a subgroup of H and $H' \backslash X = H \backslash X$ then $H' = H$.

The proof will be carried out in three parts. In Section 2 we prove the equivalence of (F) and (F'), also explaining the terminology of (F'). In Section 3 we show that (DR) implies that H is discrete, as well as the final assertions about H in the theorem. In Section 4 we deduce (DR) from the discreteness of H when H is infinite. In Section 5 we discuss how to accommodate inversions.

1.4   Minimality

We call (A, i) *minimal* if $H = G_{(A, i)}$ acts minimally on $X = \widetilde{(A, i)}$, i.e., H leaves no proper subtree invariant. This can be characterized directly in terms of (A, i) (cf. [B], (7.12) or [T], (5.7)). For example, when A is finite, (A, i) is minimal if and only if it has no vertex of (A, i)-degree 1, where, for $a \in VA, \deg_{(A, i)}(a) = \sum_{e \in E_0(a)} i(e)$.

It is easily seen that, when (A, i) is minimal, condition (DR) is equivalent to

$$(DR)_{min} \qquad [i(e) > 1] \Rightarrow \begin{cases} i(e) = 2, & \text{and} \\ A_0(e) = \{\partial_0 e\}, \end{cases}$$

Remark 1.5. Parts of Theorem (1.3) are similar in spirit to results of [T] (Corollary 6.6 and Proposition 6.7). There one considers any tree X, a map $p : VX \to A$, where A is a set, and H the group of deck transformations of p. Let H^+ denote the subgroup of H generated by edge stabilizers (called "fixators" in [T]). Under fairly general conditions ([T], Theorem 4.5) H^+ is shown to be a simple group. In our case, when H is discrete, it is easily seen (cf. loc. cit. Corollary 6.6, or (4.1) (3) below) that $H^+ = \{1\}$. Thus the structure of H given in Theorem (1.3) follows from the structure of H/H^+ given in [T], Proposition 6.7.

A similar remark applies to Theorem 5.5 below.

Corollary 1.6. *If $H = G_{(A, i)}$ is an X-lattice (a discrete automorphism group of X of finite covolume) then it is a uniform lattice, i.e., A is a finite graph.*

Proof. Let $\mathbb{A} = (A, \mathcal{A}) = H \setminus\setminus X$. Then (cf. [BL], (3.2))

$$\text{Vol}(H \setminus\setminus X) = \sum_{a \in VA} 1/|\mathcal{A}_a|.$$

Either of the conditions (b) (F) or (b) (DR) implies that $|\mathcal{A}_a|$ takes only finitely many possible values. Therefore the finiteness of $\text{Vol}(H \setminus\setminus X)$ implies that of VA, so A, being locally finite, is finite.

2   The finite case

Consider a group H acting without inversions on a locally finite tree X with quotient

$$p : X \to A = H \setminus X, \quad \text{and } (A, i) = I(H \setminus\setminus X).$$

Let $x \in VX$ and $a = p(x)$.

Proposition 2.1. *The group H fixes x if and only if A is a tree and a is a "dominant root of (A, i)," i.e., $i(e) = 1$ for all $e \in EA$ directed toward a.*

If H fixes x, then it preserves distance from x, and, along rays from x in X, p is injective and stabilizers decrease. Conversely, if (A, i, a) is a dominant rooted indexed tree, then it is clear (cf. [B], proof of (7.12)) that, for the grouping $H \setminus\setminus X = (A, \mathcal{A})$ of (A, i), we have $H \cong \pi_1(H \setminus\setminus X, a) = \mathcal{A}_a$, and so $H_x = H$.

Proposition 2.2. *If H is finite, then H fixes some $x \in VX$.*

This is well known (cf. [T] or [S] Chap. I, Prop. 19).

(2.3) With the notation of (1.3), (F) $\Rightarrow$ (F').
    Assume that H is finite. By (2.2) H fixes some $x \in VX$. Put $a = p(x)$. By (2.1), (A, i, a) is a dominant rooted edge-indexed tree. It remains to show that only finitely many edges of (A, i) are ramified.
    Let $(e_1, e_2, \ldots, e_n, \ldots)$ be a geodesic edge-path (i.e., $e_{i+1} \neq \bar{e}_i$ for all i) from $a = \partial_0 e_1$ in A. Lift it to a geodesic edge-path $(f_1, \ldots, f_n, \ldots)$ from x in X, say with vertex sequence $x_0 = x = \partial_0 f_1$, $x_1 = \partial_1 f_1 = \partial_0 f_2$, $x_2 = \partial_1 f_2 = \partial_0 f_3, \ldots$. The dominant root condition says that $i(\bar{e}_r) = 1$ (for all r), and so $H_{f_r} = H_{x_r}$, and $H_{x_{r-1}}$ contains $H_{f_r} = H_{x_r}$. Since H is finite the inclusion of H_{x_r} in $H_{x_{r-1}}$ can be strict only finitely often. It follows that only finitely many edges of the geodesic ray $(e_1, e_2, e_3, \ldots)$ are ramified. Since the tree A is locally finite, it follows further that only finitely many edges of (A, i) are ramified.

(2.4) With the notation of (1.3), (F') $\Rightarrow$ (F).
    Assume that (A, i, a) is a dominant rooted edge-indexed tree with only finitely many ramified edges, and take $x \in p^{-1}(a)$. Let $H = G_{(A,i)}$. By (2.1), H fixes x. Let

B be a ball in X centered at x and large enough so that $p(B)$ contains all ramified edges of (A, i). Then B is H-invariant and finite. Put

$$H' = \ker(H \to H|B),$$

a finite index subgroup of H. Since (A, i) is unramified outside of $p(B)$ it follows that stabilizers in H are constant along rays moving away from B. Since H' acts trivially on B it must therefore act trivially on X. But $H = G_{(A, i)}$ is a subgroup of $\mathrm{Aut}(X)$. Hence $H' = \{1\}$, so H is finite.

3 Consequences of discrete ramification: Discreteness of H and the final assertions of Theorem 1.3

(3.1) Let X be a locally finite tree. Let the subgroup H of $G = \mathrm{Aut}(X)$ act without inversion, and with quotient

$$p : X \to A = H \setminus X,$$

and set $(A, i) = I(H \backslash\backslash X)$. We assume here that

(DR) (A, i) is discretely ramified

(cf. 1.2)). Put
$$E' = \{e \in EA | i(e) = 2\},$$

the set of ramified edges of (A, i).

3.2   Case $E' = \emptyset$.

Then (A, i) is unramified. It follows that H acts freely on X, and $H = \pi_1(A) = G_{(A, i)}$, a free group (cf. [B], (1.5), Examples).

3.3   Case $(A, i) =$

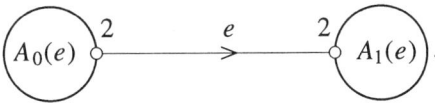

Here, by (1.2), $E' = \{e, \bar{e}\}$ and each $(A_j(e), i)(j = 0, 1)$ is an unramified tree. This (A, i) clearly admits a unique faithful grouping $\mathbb{A} = (A, \mathcal{A})$ (cf. [B], (1.24)), which consists of a constant group $\langle s_e \rangle$ of order 2 on $A_0(e)$, another, $\langle s_{\bar{e}} \rangle$, on $A_1(e)$, and the trivial group on e and $\bar{e}$. Then we have $H = \pi_1(\mathbb{A}) = \langle s_e \rangle * \langle s_{\bar{e}} \rangle$, an infinite dihedral group. The uniqueness implies that $H = G_{(A, i)}$, and H is discrete since $\mathbb{A}$ is a graph of finite groups.

3.4   Remaining cases.

For $e \in E'$, we must now, with case (3.3) aside, have $i(\bar{e}) = 1$. Put $A_1^+(e) = A_1(e) \cup \{e, \bar{e}, \partial_0 e\}$.

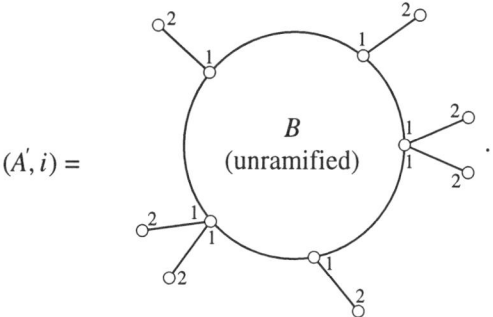

By (DR), $(A_0(e), i)$ is an unramified tree, and so $E' \subset A_1^+(e)$. Put

$$B = \bigcap_{e \in E'} A_1(e), \quad \text{and}$$

$$A' = \bigcap_{e \in E'} A_1^+(e) = B \amalg (E' \cup \bar{E}' \cup \partial_0 E').$$

Suppose for instance that,

$$(A', i) =$$

[figure: B (unramified) as a loop with branches]

Let $\mathbb{A} = (A, \mathcal{A})$ be a faithful grouping of (A, i). Since (B, i) is unramified, $\mathcal{A}|B$ is a constant group, H_B. Similarly, for $e \in E'$, $(A_0(e), i)$ is unramified, so $\mathcal{A}|A_0(e)$ is a constant group, $H_{(e)}$. Moreover α_e embeds $H_B = \mathcal{A}_{\partial_1 e} = \mathcal{A}_e$ into $H_{(e)}$ as an index 2, hence normal, subgroup. Thus H_B defines a normal subgroup of $\pi_1(\mathbb{A})$ that lies in every vertex and edge stabilizer, and so acts trivially on X. Since the grouping was taken to be faithful, $H_B = \{1\}$. Hence $H_{(e)} = \langle s_e \rangle$, a group of order 2. This shows that the faithful grouping $\mathbb{A}$ of (A, i) is unique and finite, hence $H = \pi_1(\mathbb{A}) = G_{(A, i)}$, H is discrete and

$$H \cong \left(\underset{e \in E'}{*} \langle s_e \rangle \right) * \underbrace{\pi_1(A)}_{=\pi_1(B)}.$$

4 $G_{(A, i)}$ discrete and infinite implies (A, i) discretely ramified

Here we assume that $H = G_{(A, i)}$ is discrete and infinite, and we propose to show that (A, i) is discretely ramified (DR). Let $\ell : H \to \mathbb{Z}$ denote the hyperbolic length function of the action of H on $X = \widetilde{(A, i)}$ (see [B], (6.2)).

4.1   The case $\ell(\mathbf{H}) \neq 0$

In this case there is a unique minimal H-invariant subtree $Y = X_H$ of X (union of the translation axes of the hyperbolic elements in H) ([B], (7.5)). Let $r : VX \to VY$ be the "geodesic retraction," defined by the condition that $[x, r(x)] = [x, Y]$, the geodesic path from $x \in VX$ to Y. Then, for $y \in VY, r^{-1}(y) = VX_y$, where (X_y, y) is a rooted tree, called the "normal tree to Y at y in X."

Consider the exact sequence

$$1 \to N \to H \xrightarrow{\text{res}} H|Y \to 1,$$

where N consists of the elements of H fixing the vertices of Y. It follows from the fact that $H = G_{(A, i)}$ that

$$N = \prod_{y \in VY} (H_y | X_y).$$

(1) **Claim:** $N = \{1\}$.

If not there is a $y \in VY$ such that $(H_y|X_y) \neq \{1\}$. For $h \in H$, $(H_{hy}|X_{hy}) = h(H_y|X_y)h^{-1} \neq \{1\}$. Since H is discrete, the group N, being contained in H_y, is finite. Hence $\prod_{z \in H \cdot y}(H_z|X_z)(\cong (H_y|X_y)^{|H \cdot y|})$ is finite, and so the orbit $H \cdot y$ must be finite. But then (cf. [S], Chap. I, Prop. 1.9) $\ell(H) = 0$, contrary to assumption. Whence the claim.

Consider

$$B = H \backslash Y \subset A = H \backslash X.$$

From Claim (1) it follows that $H_y = H_z$ for all $y \in VY$ and $z \in VX_y$, and so

(2) $i(e) = 1 \quad$ for all $e \in EA - EB$

(3) **Claim:** $H_e = \{1\}$ for all $e \in EY$.

Suppose, on the contrary, that $e \in EY$, $h \in H_e$, and $h \neq 1$.

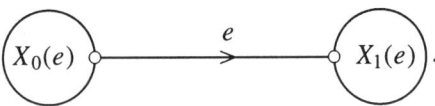

Since $H = G_{(A, i)}$ we have $H_e = (H_e|X_0(e)) \times (H_e|X_1(e))$. Hence we can choose h above so that, for example, $h|X_0(e) \neq 1$ but $h|X_1(e) = 1$.

Since $Y = X_H$ is a union of hyperbolic axes X_g of hyperbolic elements g of H, we can choose such a g so that $e \in EX_g$.

Let Z be any large ball in X centered at a vertex of X_g. Since Z is finite we can choose an $M \in \mathbb{Z}$ so that $g^M Z \subset X_1(e)$. Thus $h|g^M Z = 1$, and so $g^{-M}hg^M|Z = 1$, with $1 \neq g^{-M}hg^M \in H$. Since Z was an arbitrarily large ball, this shows that H is not discrete, contrary to hypothesis. Hence $H_e = \{1\}$, as claimed.

(4) **Claim:** If $y \in VY$ and $H_y \neq \{1\}$, then

$$|H_y| = 2 = \deg_Y(y).$$

Indeed, it follows from (3) that H_y acts freely on $E_0^Y(y)$. Since $H = G_{(A, i)}$ it follows that $H_y | E_0^Y(y)$ is the direct product of the full symmetric groups on each of the H_y-orbits on $E_0^Y(y)$. This action can be free, with $H_y \neq \{1\}$, only if $E_0^Y(y)$ consists of a single 2-edge orbit, and $|H_y| = 2$.

(5) (A, i) is (DR) (notation of (1.2)).

Say $e \in EA$ and $i(e) > 1$. From (2) it follows that $e \in EB$, where, recall, $B = H \setminus Y$. Let $a = \partial_0 e$, $y \in p^{-1}(a)(\subset VY)$, and $f \in p_{(y)}^{-1}(e) \subset E_0^Y(y)$. Then

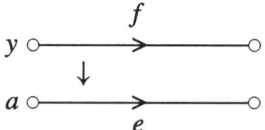

$[H_y : H_f] = i(e) > 1$. It follows then from (4) that $|H_y| = 2 = \deg_Y(y)$. Hence (B, i) has the form

$$(B, i) = \underset{i(e) = 2}{\circ} \overset{a}{\underset{}{\xrightarrow{\hspace{2cm}}}} \overset{e}{} \circ \boxed{B_1(e)} .$$

Moreover, we see from (1) and (2) that

$$(A, i) = \boxed{(A_0(e), i)}_2 \overset{e}{\xrightarrow{\hspace{3cm}}} \circ \boxed{(A_1(e), i)} .$$

where $p : X_y \to A_0(e)$ is an isomorphism (hence $A_0(e)$ is a tree) and $(A_0(e), i)$ is unramified. This proves (5), as required.

4.2   The case $\ell(H) = 0$

If H fixes a vertex then, by discreteness, H is finite, contrary to assumption. Hence, since H acts without inversion, H fixes a unique end, ε, but no vertex (see [B], (7.2)). We conclude the proof of Theorem (1.3) by showing that this leads to a contradiction.

For $x \in VX$, H_x fixes the ray $[x, \varepsilon) =$

$$x = x_0 \quad\;\; x_1 \quad\;\; x_2 \quad\;\; x_3 \quad\;\; x_4$$
$$\circ \xrightarrow{} \circ \xrightarrow{} \circ \xrightarrow{} \circ \xrightarrow{} \circ \cdots$$
$$e_1 \qquad\; e_2 \qquad\; e_3 \qquad\; e_4$$

We can view X as obtained by attaching a rooted tree $(Y_{x,\varepsilon}(x_n), x_n)$ to x_n, for each $n \geq 0$. Since $H = G_{(A,\,i)}$ we have

$$H_x = \prod_{n \geq 0} (H_x | Y_{x,\varepsilon}(x_n)).$$

Since H_x (by discreteness) is finite we have $H_x | Y_{x,\varepsilon}(x_n) = \{1\}$ for all but finitely many n. We shall derive a contradiction by showing that $H_x | Y_{x,\varepsilon}(x_n) \neq \{1\}$ for infinitely many n.

The group H is the union of the increasing sequence of subgroups $(H_{x_0}, H_{x_1}, \dots)$. Since H fixes no vertex we have $H_{x_{n-1}} \neq H_{x_n}$ infinitely often. Put

$$q_n = [H_{x_n} : H_{x_{n-1}}] = |H_{x_n} \cdot \bar{e}_n|.$$

Since $H = G_{(A,\,i)}$, H_{x_n} acts on $H_{x_n} \cdot \bar{e}_n$ as a full symmetric group, S_{q_n}. If $q_n > 2$ then H_{x_n} has non-trivial elements fixing $\bar{e}_n$, and hence acting non-trivially on $Y_{x,\varepsilon}(x_n)$. Thus, if $q_n > 2$ infinitely often, we have the desired contradiction.

So assume now that $q_n \leq 2$ for all but finitely many n, and $q_n = 2$ infinitely often. Replacing x by some x_n, we can further arrange that $q_n \leq 2$ for all n, and $q_1 = 2$. Put $Y_n = Y_{x,\varepsilon}(x_n)$.

Say $H_{x_1} \cdot e_1 = \{e_1, f_1\}$. Since $H = G_{(A,\,i)}$ there is an $h \in H_{x_1}$ such that $he_1 = f_1$, $hf_1 = e_1$, and $h|Y_n = 1$ for all $n \geq 2$:

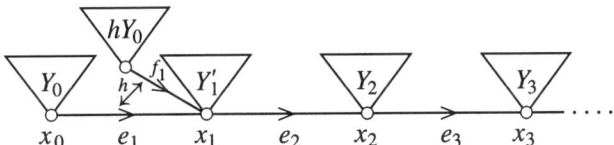

Choose $n > 1$ so that $q_n = 2$ and $h_n \in H_{x_n}$ such that $h_n e_n = f_n \neq e_n$:

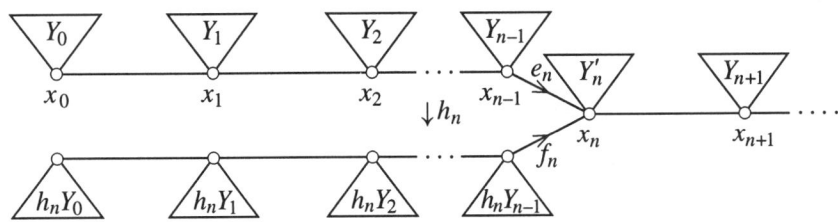

Then we have $1 \neq h_n h h_n^{-1} \in H_{x_n}$, and $h_n h h_n^{-1}$ acts non-trivially on

$$h_n \left(\begin{array}{c} \overset{Y_0}{\diagdown} \quad \overset{Y_1}{\diagdown} \\ x_0 \qquad x_1 \end{array} \right) = \diagup h_n Y_0 \diagdown \quad \diagup h_n Y_1 \diagdown \subset Y_n.$$

Thus $H_{x_n} | Y_n \neq \{1\}$, and this happens for infinitely many n; contradiction.

5   Accommodating inversions

(5.1) Let X be locally finite tree, $G = \text{Aut}(X)$, and H, a subgroup of G with quotient

$$p : X \to A = H \setminus X.$$

If H acts with inversions, say $s \in H$ inverts $e \in EX$, $se = \bar{e}$, then $f = p(e) \in EA$ is a "self-inverse loop," i.e., $f = \bar{f}$. Thus A is not a graph in our sense, but a "quasi-graph" (or "q-graph") in the sense that we allow the presence of self-inverse loops in EA. When $\bar{f} = f$, we put

$$A_0(f) = A - \{f\}, \qquad \left(A_0(f) \bigcirc \right) f = \bar{f}.$$

We can still define the edge-indexed q-graph

$$(A, i) = I(H \setminus\setminus X)$$

as previously: If $e \in E_0^X(x)$ and $f = p(e) \in E_0^A(p(x))$, then $i(f) = |p_{(x)}^{-1}(f)| = |H_x \cdot e| = [H_x : H_e]$.

(5.2) Let

$$E' = \{e \in EX | se = \bar{e} \quad \text{for some } s \in H\}.$$

Clearly, $\overline{E'} = E'$. Let X' denote the tree obtained from X by barycentrically subdividing the edges in E'. Each

$$\begin{array}{ccc} x_0 & e & x_1 \\ \circ & \!\!\!\!\!\longrightarrow\!\!\!\!\! & \circ \end{array} \quad \text{in } E'$$

is replaced in X' by

$$\begin{array}{ccccc} x_0 & e_0 & m(e) & e_1 & x_1 \\ \circ & \!\!\longrightarrow\!\! & \circ & \!\!\longrightarrow\!\! & \circ \end{array}$$

with $\partial_0 e_0 = x_0 = \partial_0 e$, $\partial_1 e_1 = x_1 = \partial_1 e$, $\partial_1 e_0 = m(e) = \partial_0 e_1$, and $\overline{(e_j)} = (\bar{e})_{1-j} (j = 0, 1)$ (so $m(e) = m(\bar{e})$).

Then H naturally acts without inversions on X'. Moreover,

$$(A', i') = I(H \setminus\setminus X')$$

is obtained from (A, i) as follows. If $f = \bar{f} \in EA$ is a self-inverse loop and $i(f) = r$, then

$$a \left(\bigcirc i(f) = r \right) f \quad \text{in } (A, i)$$

is replaced by

$$a \underset{i'(f') = r = i(f)}{\overset{f'}{\circ \longrightarrow}} \underset{i'(\bar{f}') = 2}{\circ} m(f) \quad \text{in } (A', i'),$$

where $\partial_0 f' = a = \partial_0 f (= \partial_1 f)$, $\partial_1 f' = m(f)$ (a new vertex), $i'(f') = r = i(f)$, and $i'(\bar{f}') = 2$.

It is clear that we can identify the subgroup

$$K = G_{(A, i)} = \{g \in G | p \circ g = p\}$$

of G with the subgroup of $G' = \text{Aut}(X')$,

$$K' = G'_{(A', i')} = \{g \in G' | p' \circ g = p'\},$$

where $p' : X' \to A' = H \setminus X'$. For $e \in EX$, $\partial_0 e = x$, we have $K_x = K'_x$, and $K_e = K'_e$ if $e \notin E'$. If $e \in E'$ then $K_e = K'_{e_0} = K'_{e_1}$, and $K'_{m(e)}$ is the stabilizer, in K, of $\{e, \bar{e}\}$. In particular, K is discrete if and only if K' is discrete.

(5.3) From the discussion above we see that (A', i') satisfies (DR) if and only if (A, i) satisfies the following extension of the notion of (DR) to edge-indexed q-graphs:

$$(DR)_q \qquad [i(e) > 1] \Rightarrow \begin{cases} i(e) = 2, \\ e \text{ is a self inverse loop or a separating edge,} \\ (A_0(e), i) \text{ is an unramified edge-indexed tree.} \end{cases}$$

(5.4) We further conclude that the finiteness of K, equivalent to condition (F') of (1.3) for (A', i'), is equivalent to the following condition for (A, i).

$(F')_q$ For some $a \in VA$, either (A, i, a) is a dominant-rooted edge-indexed tree with only finitely many ramified edges, or else it is such an object with a single unramified self-inverse loop attached at a.

With these observations, Theorem (1.3) now takes the general form.

Theorem 5.5. *Let X be a locally finite tree, H a subgroup of $G = \text{Aut}(X)$, $p : X \to A = H \setminus X$, and $(A, i) = I(H \setminus X)$. Assume that*

$$H = G_{(A, i)} = \{g \in G | p \circ g = p\}.$$

The following conditions are equivalent.

(a) *H is discrete.*

(b) *Either*

    (F) *H is finite*

or

(DR)$_q$ (A, i) *is discretely ramified (in the sense of* (5.3)).

Further, condition (b)(F) *is equivalent to condition* (F')$_q$ *of* (5.4).
Finally, (DR)$_q$ *implies that*

$$H \cong \left(\underset{\substack{e \in EA' \\ i(e)=2}}{*} \langle s_e \rangle \right) * \pi_1(A)'.$$

where $\pi_1(A')$ *is a free group, each* s_e *has order* 2, *and* (A', i') *is obtained from* (A, i) *as in* (5.6) *below.*

(5.6) Let (A, i) be any edge-indexed q-graph. Let $L = L(A) = \{e \in EA | \bar{e} = e\}$, the set of self-inverse loops of A. We define the *associated edge-indexed graph*

$$(A', e') = \beta(A, i)$$

as follows. For $e \in L$, $\partial_0 e = a(= \partial_1 e)$,

$$a \,\,\bigcirc\!\!i(e)\,\, e = \bar{e} \quad \text{in } (A, i)$$

is replaced by

$$a \circ \xrightarrow[\,i'(\beta e) = i(e)\,]{\beta e} \xrightarrow[\,i'(\overline{\beta e}) = 2\,]{} \circ\, m(e) \quad \text{in } (A', i').$$

More precisely, there are injections $m : L \to VA'$ and $\beta : EA \to EA'$ such that $VA' = VA \amalg m(L)$ and $EA' = \beta EA \amalg \overline{\beta L}$. For $e \in EA - L$, $\partial_h(\beta e) = \partial_h e (h = 0, 1)$, $\overline{\beta e} = \beta(\bar{e})$, and $i(\beta e) = i(e)$. For $e \in L$, $\partial_0 \beta e = \partial_0 e$, $\partial_1 \beta e = m(e)$, and $i'(\beta e) = i(e)$ and $i'(\overline{\beta e}) = 2$.

Let $p' : X' = \widehat{(A', i')} \to A'$ denote the universal covering tree, $G' = \text{Aut}(X')$, and

$$G'_{(A', i')} = \{g \in G' | p' \circ g = p'\}.$$

Since (A', i') admits an effective grouping, e.g., the canonical cyclic grouping ([BK], (2.2)) it follows (cf. [B]), that $(A', i') = I(G'_{(A',i')} \backslash\!\backslash X')$.

Let $M = p'^{-1}(m(L)) \subset VX'$. This is a $G'_{(A',i')}$-invariant set of vertices of degree 2. Let X be the tree obtained by removing M, and replacing the two geometric edges incident to $y \in M$ with a single edge $e(y)$,

$$x \circ\!\!\xrightarrow{}\!\! y \circ\!\!\xrightarrow{}\!\! z \circ \quad \text{in } X' \quad \mapsto \quad x_0 \circ \xrightarrow{e(y)} x_1 \circ \quad \text{in } X.$$

We call X the universal covering tree of (A, i), and denote it by

$$X = \widetilde{(A, i)}.$$

To justify this consider $G = \mathrm{Aut}(X)$. The action of $G'_{(A', i')}$ on X' naturally defines an action also on X, and hence a subgroup $G_{(A, i)}$ of G. For $y \in M$, the stabilizer in $G'_{(A', i')}$ of y corresponds to a subgroup of $G_{(A, i)}$ that inverts the edge $e(y)$ of X. It is easy to see from this picture that we can identify (A, i) with $I(G_{(A, i)} \backslash\backslash X)$. In particular we have $p : X \to A = G_{(A, i)} \backslash X$, and it is further clear that $G_{(A, i)} = \{g \in G | p \circ g = p\}$.

A more direct combinatorial construction of $X = \widetilde{(A, i)}$ can be given exactly following the description in [B], Remark (1.18), in terms of reduced paths.

Let (B, j) be another edge-indexed q-graph and $q : A \to B$ a q-graph morphism. We call $q : (A, i) \to (B, j)$ a *covering* if, for all $a \in VA$, $b = q(a)$, and $f \in E_0^B(b)$, we have

$$j(f) = \sum_{e \in q_{(a)}^{-1}(f)} i(e).$$

Under this condition it follows just as in [B] (2.8), that we can identify $X = \widetilde{(A, i)}$ with $\widetilde{(B, j)}$ so that we have a natural commutative diagram of projections

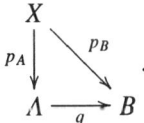

6   The ultimate quotient $(A_*, i_*) = I(G \backslash\backslash X)$

(6.1) Equivalence relations. Let V be a set. The set $Eq(V)$ of equivalence relations on V, viewed as subsets of $V \times V$, is ordered by inclusion. The diagonal is the minimal element, and defines equality. The maximal element, $V \times V$, defines the "egalitarian relation," all members are equivalent.

For $R \in Eq(V)$ and $x \in V$, let x_R denote the R-class of x, and $p_R : V \to V/R$ the quotient, $p_R(x) = x_R$.

If $f : V \to W$ is a set map it defines $R(f) \in Eq(V)$ by $(x, y) \in R(f)$ if and only if $f(x) = f(y)$. It also defines $f^{-1} : Eq(W) \to Eq(V)$ by $f^{-1}R = R(p_R \circ f)$, i.e., $(x, y) \in f^{-1}R$ if and only if $(f(x), f(y)) \in R$. When f is surjective we can identify $p_{f^{-1}R}$ with $p_R \circ f$.

(6.2) Edge connections and degrees. Let A be a q-graph (see (5.1)). For $C, D \subset VA$

put

$$E(C, D) = \{e \in EA | \partial_0 e \in C \quad \text{and} \quad \partial_1 e \in D\}$$
$$= \overline{E(D, C)}.$$

Now let (A, i) be a locally finite edge-indexed q-graph. Then, for $C, D \subset VA$, we put

$$i(C, D) = \sum_{e \in E(C,D)} i(e).$$

When $C = \{a\}$, $a \in VA$, we also write

$$i(a, D) = i(\{a\}, D).$$

Thus

$$i(a, D) \leq i(a, VA) = \deg_{(A, i)}(a)$$

for all $a \in VA$ and $D \subset VA$.

(6.3) The degree refinement operator ρ. (This construction generalizes that in [L].) Let (A, i) be an edge indexed q-graph. *Degree refinement* refers to the operator

(1) $$\rho : Eq(VA) \to Eq(VA),$$

defined on $R \in Eq(VA)$ as follows.

(2) $(a, b) \in \rho R \Leftrightarrow (a, b) \in R$,   and $i(a, c_R) = i(b, c_R)$ for $c \in VA$.

Note that

(3) $$\rho R \subset R, \quad \text{and} \quad [R \subset R' \Rightarrow \rho R \subset \rho R'],$$

the latter because R'-classes are unions of R-classes.

Suppose that a group H acts on A preserving i: $i(he) = i(e)$ for $e \in EA$ and $h \in H$. Clearly,

(4)   *if H preserves R-classes ($ha_R = a_R$ for $a \in VA$) then H preserves ρR-classes.*

(5)   *Let $q : (A, i) \to (A', i')$ be a covering of edge indexed q-graphs, $R' \in Eq(VA')$, and $R = q^{-1}R'$. Then $\rho R = q^{-1}(\rho R')$.*

Proof. Let $a, b \in VA$. Then $(a, b) \in \rho R$ if and only if (i) $(a, b) \in R$ and (ii) $i(a, c_R) = i(b, c_R)$ for $c \in VA$. Since $R = q^{-1}R'$, (i) $\Leftrightarrow$ (i') : $(qa, qb) \in R'$. We claim that

(*)   (ii) $\Leftrightarrow$ (ii') : $i'(qa, (qc)_{R'}) = i'(qb, (qc)_{R'})$ for $c \in VA$.

Since $(a, b) \in q^{-1}(\rho R')$ if and only if $(qa, qb) \in \rho R'$, if and only if [(i') and (ii')], (5) will follow from $(*)$. In turn $(*)$ clearly follows from

$$(**) \qquad\qquad i'(qa, (qc)_{R'}) = i(a, c_R) \; \forall a, c \in VA.$$

To see this first note that $q^{-1}(qc)_{R'} = c_R$. In fact $b \in q^{-1}(qc)_{R'} \Leftrightarrow qb \in (qc)_{R'} \Leftrightarrow (qb, qc) \in R' \Leftrightarrow (b, c) \in R(= q^{-1}R') \Leftrightarrow b \in c_R$. Now we have

$$
\begin{aligned}
i'(qa, (qc)_{R'}) &= \sum_{e' \in E(qa, (qc)_{R'})} i'(e') \\
&= \sum_{e' \in E(qa, (qc)_{R'})} \sum_{e \in q_{(a)}^{-1}(e')} i(e) \qquad \text{(since q is a covering)} \\
&= \sum_{e \in E(a, q^{-1}(qc)_{R'})} i(e) \\
&= \sum_{e \in E(a, c_R)} i(e) \qquad\qquad (q^{-1}(qc)_{R'} = c_R) \\
&= i(a, c_R).
\end{aligned}
$$

This completes the proof of $(**)$, and so also of (5).

(6.4) Fixed points of ρ. Let (A, i) be an edge-indexed q-graph. Let $R \in Eq(VA)$ be a fixed point of ρ, $\rho R = R$, with projection $p : VA \to VA/R$, $p(a) = a_R$.
   For $a, b \in VA$ define

$$(1) \qquad\qquad i'(a, b) = i(a, b_R).$$

If $a_1 \in a_R$ then $(a, a_1) \in R = \rho R$, and so $i(a_1, b_R) = i(a, b_R)$. Thus $i'(a, b)$ depends only on a_R and b_R so we can unambiguously define

$$(1') \qquad\qquad i'(a_R, b_R) = i(a, b_R) \text{ for } a, b \in VA.$$

We now define a q-graph A' by

$$(2) \qquad
\begin{cases}
VA' = VA/R, \\
EA' = \{(a', b') \in VA' \times VA' | i'(a', b') > 0\} \\
\partial_0(a', b') = a', \quad \partial_1(a', b') = b', \quad \text{and} \\
\overline{(a', b')} = (b', a').
\end{cases}
$$

For the last item (inversion) we must check that $i'(a', b') > 0$ implies $i'(b', a') > 0$. Say $a' = a_R$ and $b' = b_R$. By hypothesis, there is an $e \in E(a, b_R)$. Say $\partial_1 e = b_1 \in b_R$. Then $\bar{e} \in E(b_1, a_R)$ and so $0 < i(b_1, a_R) = i'(b_{1R}, a_R) = i'(b', a')$.
   We use the notation

$$(3) \qquad\qquad A' = A/R \quad \text{and} \quad (A', i') = (A, i)/R.$$

We further have the q-graph morphism

(4)
$$p : A \to A', \quad p(a) = a_R \quad \text{for } a \in VA,$$
$$\text{and} \quad p(e) = ((\partial_0 e)_R, (\partial_1 e)_R) \quad \text{for } e \in EA.$$

We claim that

(5) $p : (A, i) \to (A', i')$ is a covering of edge-indexed q-graphs, i.e., for $(a_R, b_R) \in EA'$

$$i'(a_R, b_R) = \sum_{e \in p_{(a)}^{-1}((a_R, b_R))} i(e).$$

In fact, $e \in p_{(a)}^{-1}(a_R, b_R)$ if and only if $\partial_0 e = a$ and $\partial_1 e \in b_R$, i.e., if and only if $e \in E(a, b_R)$. Thus (5) is just the relation (1′).

It follows from (4) and (5) that we have a common universal covering tree

(6)
$$\widetilde{(A, i)} = X = \widetilde{(A', i')}$$

and a commutative diagram of projections

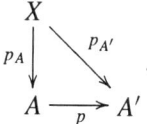

By construction, $(A', i') = (A, i)/R$ has the following property:

(7)   *For all $a', b' \in VA'$, there is at most one edge of A' from a' to b'. In particular, every loop of A' is self-inverse.*

(6.5) The relations $R_N = R_N(A, i) \in Eq(VA)$. Starting with the "egalitarian relation,"

(0)
$$R_0 = VA \times VA,$$

we define R_N inductively by

(1)
$$R_{N+1} = \rho R_N \subset R_N, \quad \text{for } N \geq 0,$$

and put

(2)
$$R_* = R_*(A, i) = \bigcap_{N \geq 0} R_N.$$

Thus $(a, b) \in R_0$ for all $a, b \in VA$, and

(3)
$$(a, b) \in R_1 \Leftrightarrow \deg_{(A, i)}(a) = \deg_{(A, i)}(b).$$

We shall use the following abbreviated notation for $N \in \mathbb{N} \cup \{*\}$:

(4) $$p_N = p_{R_N} : VA \to VA/R_N, \ p_N(a) = a_N(= a_{R_N}),$$

(5) $$p_{a,N} : E_0(a) \to VA/R_N, \ p_{a,N}(e) = (\partial_1 e)_N.$$

For $c \in VA$ we clearly have

(6) $$i(a, c_N) = \sum_{e \in p_{a,N}^{-1}(c_N)} i(e),$$

which shows that the function $i(a, -) : VA/R_N \to \mathbb{N}$ depends only on the fibers of $p_{a,N}$.

It further follows inductively from (6.3)(4) that

(7) For $a \in VA$ and $N \in \mathbb{N} \cup \{*\}$, a_N is Aut(A, i)-invariant.

Proposition 6.6.
  (a) $\rho R_* = R_*$.
  (b) If $R \in Eq(VA)$ and $\rho R = R$, then $R \subset R_*$.
  (c) $R_* = R_N$ if $N \geq |VA|$.

Assertion (c) is obvious. To prove (b) observe that $R \subset R_0$, so $R = \rho^N R \subset \rho^N R_0 = R_N$ for all $N \in \mathbb{N}$. We now prove (a). For $a \in VA$ and $N \in \mathbb{N}$ we have a commutative diagram,

$$E_0(a) \xrightarrow{p_{a,N}} VA/R_N$$

$$\searrow_{p_{a,*}} \qquad \downarrow$$

$$VA/R_*.$$

Since $E_0(a)$ is finite, $p_{a,N}$ and $p_{a,*}$ have the same fibers for $N \gg 0$, and hence, for $N \gg 0$,

$$i(a, c_N) = i(a, c_*) \quad \text{for } c \in VA.$$

To show that $\rho R_* = R_*$ we must show that if $(a, b) \in R_*$ then $(a, b) \in \rho R_*$, i.e., that $i(a, c_*) = i(b, c_*)$ for $c \in VA$. For $N \gg 0$ we have seen that $i(a, c_*) = i(a, c_N)$ and $i(b, c_*) = i(b, c_N)$. Since $(a, b) \in R_* \subset R_{N+1} \subset R_N$ we have $i(a, c_N) = i(b, c_N)$, whence $i(a, c_*) = i(b, c_*)$, as required.

Example 6.7. Let

$$A = \overset{0}{\circ} \text{——} \overset{1}{\circ} \text{——} \overset{2}{\circ} \text{——} \overset{3}{\circ} \text{——} \overset{4}{\circ} \cdots,$$

the infinite half-line, as an unramified edge-indexed graph. For $N \in \mathbb{N} = VA$ put $[N] = \{N, N + 1, N + 2, \dots\}$. With this notation, the R_N-classes in VA are easily seen to be $\{0\}, \{1\}, \dots, \{n - 1\}, [N]$. In particular R_* is the identity relation, and $R_* \neq R_N$ for $N \in \mathbb{N}$.

Following are some interesting properties of the relations R_N.

Lemma 6.8. *For $a, b \in VA$ let $d(a, b)$ denote the (minimum edge-path) distance from a to b in A. For $0 \le m \le N$, and $a, b, c \in VA$,*

$$\left. \begin{array}{r} a_N = b_N \\ d(a, c_m) \le N - m \end{array} \right\} \Rightarrow d(a, c_m) = d(b, c_m).$$

Illustration ($m = 1$): Suppose that $a_N = b_N$ and c is the nearest vertex to a such that $\deg_{(A, i)}(c) = v$. If $d(a, c) < N$ and if c' is the nearest vertex to b such that $\deg_{(A, i)}(c') = v$, then $d(b, c') = d(a, c)$.

Proof of (6.8). We argue by induction on $N - m$. It suffices, by symmetry, to show (assuming that $a_N = b_N$) that

$$d(a, c_m) \le N - m \Rightarrow d(b, c_m) \le N - m.$$

Case $N - m = 0$. $d(a, c_N) = 0 \Rightarrow c_N = a_N = b_N \Rightarrow d(b, c_N) = 0$.

Case $N - m = 1$. Say $d(a, c_{N-1}) \le 1$. If $d(a, c_{N-1}) = 0$ then $c_{N-1} = a_{N-1} \supset a_N = b_N$, which contains b, so $d(b, c_{N-1}) = 0$. So assume that $d(a, c_{N-1}) = 1$. Then clearly $i(a, c_{N-1}) > 0$. Since $a_N = b_N$ we have $i(b, c_{N-1}) = i(a, c_{N-1}) > 0$, whence $d(b, c_{N-1}) \le 1$.

Case $N - m > 1$. Choose $a' \in VA$ and $c' \in c_m$ such that $d(a, a') \le N - m - 1 = N - (m + 1)$ and $d(a', c') = 1$,

We have $d(a, a'_{m+1}) \le d(a, a') \le N - (m + 1)$, so, by induction, $d(b, a'_{m+1}) \le N - (m+1)$. Choose $b' \in a'_{m+1}$ so that $d(b, b') \le N - (m+1)$. We have $a'_{m+1} = b'_{m+1}$ and $d(a', c_m) \le d(a', c') = 1 = (m + 1) - m$. By the case $N - m = 1$, we have $d(b', c_m) \le 1$. Thus $d(b, c_m) \le d(b, b') + d(b', c_m) \le N - (m + 1) + 1 = N - m$, as was to be shown.

(6.9) It follows by induction from (6.3)(5) that:

If $q : (A, i) \to (A', i')$ is a covering of edge-indexed q-graphs then

$$R_N(A, i) = q^{-1} R_N(A', i')$$

for $N \in \mathbb{N} \cup \{*\}$, and hence $VA/R_N(A, i) \overset{\cong}{\to} VA'/R_N(A, i')$.

Proposition 6.10. *Suppose that $\deg_{(A, i)}(a) \le d$ for all $a \in VA$. Then*

$$|VA/R_n| \le d^{d^{(n-1)}}$$

for all $n \ge 1$.

Proof. Thanks to (6.9) we can, if necessary, replace (A, i) by its universal covering tree, and so reduce to the case when (A, i) is unramified, which we now assume. Put $r_n = |VA/R_n|$. Then (see (6.5)) $r_0 = 1$ and $r_1 \leq d$. From (6.5)(5) we have the commutative diagrams, for $a \in VA$,

(1)
$$E_0(a) \xrightarrow{p_{a,n-1}} VA/R_{n-1}$$

with $p_{a,n}$ mapping down to VA/R_n.

From (6.5)(6) we see that a_{n+1} is determined by a_n and $p_{a,n}$. Since $p_{a,n}$ determines $p_{a,m}$ for $m \leq n$ (by (1)), it follows that $p_{a,n}$ determines a_{n+1}. The number of possible functions $p_{a,n}$ from the set $E_0(a)$ of cardinal $\leq d$, to VA/R_n is $\leq (r_n)^d$, whence $r_{n+1} \leq (r_n)^d$. Starting with $r_1 \leq d$, this gives inductively, $r_n \leq d^{d^{(n-1)}}$, as claimed.

Theorem 6.11. *Let (A, i) be an edge-indexed q-graph, $X = \widetilde{(A, i)}$, $G = Aut(X)$, $R_* = R_*(A, i)$ (cf. (6.5)), and $(A_*, i_*) = (A, i)/R_*$. Then $(A_*, i_*) = I(G \backslash\backslash X)$.*

Proof. Let $(X, 1)$ denote the unramified indexing of X. Then the projection $p_A : X \to A$ is a covering $(X, 1) \to (A, i)$. Hence, by (6.5)(5), $p_A^{-1} R_* = R_*^X := R_*(X, 1)$, and so we can identify $(A_*, i_*) = (A, i)/R_*$ with $(X, 1)/R_*^X$. Thus we are reduced to showing that the R_*^X-classes are exactly the G-orbits. It follows from (6.5)(7) that G preserves $R_N(X, 1)$-classes for all $N \in \mathbb{N} \cup \{*\}$. In particular each R_*^X-class is a union of G-orbits. Hence we have a factorization of coverings

$$(X, 1) \to (A_G, i_G) = I(G \backslash\backslash X) \to (A_*, i_*).$$

On the other hand, $(A_*, i_*) = I(H \backslash\backslash X)$, where $H = G_{(A_*, i_*)}$ is a subgroup of G. Whence the theorem.

Corollary 6.12. *Let (A, i) be an edge-indexed q-graph, $X = \widetilde{(A, i)}$, and $G = Aut(X)$. The following conditions are equivalent.*

(a) *G is discrete.*

(b) *Either G is finite, i.e., (A_*, i_*) satisfies (5.4) $(F')_q$, or else (A_*, i_*) is discretely ramified (condition $(DR)_q$ of (5.3)).*

Remark 6.13. Much of the discussion of this section can be carried out in general, without the local finiteness assumptions. For this one considers edge-indexed q-graphs (A, i) where A is any connected q-graph and the indices $i(e)$ can be arbitrary cardinals > 0. Then one can inductively define relations $R_\alpha = R_\alpha(A, i)$ for ordinals α, by $R_0 = VA \times VA$, $R_{\alpha+1} = \rho R_\alpha$, and $R_\alpha = \cap_{\beta < \alpha} R_\beta$ if α is a limit ordinal. Then, as above, one defines $R_* = R_*(A, i) = \cap_\alpha R_\alpha$, shows that $\rho R_* = R_*$, constructs $(A_*, i_*) = (A, i)/R_*$, and shows that $(A_*, i_*) = I(G \backslash\backslash X)$. The background ramified covering theory is available in this generality.

7   Examples; π-rigid graphs

7.1   Illustration of Theorem 5.5

For $r \geq 1$ consider the self-inverse indexed loop,

$$L_r = a \overset{\displaystyle i(e)}{\underset{=r}{\bigcirc}} e = \bar{e}, \quad \text{with}$$

$$\beta L_r = a \underset{r}{\circ} \xrightarrow{\quad\beta e\quad} \underset{2}{\circ} m(e).$$

Then $\widetilde{\beta L_r} = X_{r,2}$, the barycentric subdivision of the r-regular tree $X_r : \widetilde{\beta L_1} =$
$\circ\!\!\!-\!\!\!-\!\!\!\circ\!\!\!-\!\!\!-\!\!\!\circ$ and $\widetilde{\beta L_2} \cong X_2$.
Suppose that

$$(A, i) = \left((A_0(e), i) \bigcirc L_r \right) e.$$

Let $X = \widetilde{(A,\, i)}$ and $H = G_{(A,\, i)}$ in $\mathrm{Aut}(X)$. Then, by Theorem (5.5), H is finite if and only if $r = 1$ and $(A_0(e), i, a)$ is a dominant rooted edge-indexed tree with only finitely many ramified edges. If $r > 1$ then H is discrete if and only if $r = 2$ and $(A_0(e), i)$ is an unramified tree.

Proposition 7.2. *Let A be a finite connected graph, X its universal covering tree, and $G = \mathrm{Aut}(X)$. The following conditions are equivalent. When satisfied, we call A a π-rigid graph.*

(a) *The fundamental group $\pi_1(A)$ is all of G.*

(b) *For some $N \in \mathbb{N}$, $R_N(A, 1) = Id$ (notation of (6.5)). (Here $(A, 1)$ denotes A with unramified indexing.)*

(c) *We have $R_*(A, 1) = Id$ (notation of (6.5)).*

Under these conditions A is a simplicial graph (at most one edge from one vertex to another, and no loops), and $\mathrm{Aut}(A) = \{1\}$.

Since $(A, 1) = I(\pi_1(A) \backslash\backslash X)$ we see from Theorem (6.11) that (a) is equivalent to the condition that $(A, 1) = I(G \backslash\backslash X) = (A_*, 1_*) = (A, 1)/R_*(A, 1)$, whence the equivalence of (a) and (c). The equivalence of (b) and (c) follows from Proposition (6.6) (c) and the finiteness of VA.

There is always a subgroup H of G containing $\pi_1(A)$ as a normal subgroup, with quotient $H/\pi_1(A) = \mathrm{Aut}(A)$. Thus condition (a) implies that $\mathrm{Aut}(A) = \{1\}$. The latter property, in turn, easily implies that A is a simplicial graph.

If A has no terminal vertices we can "π-rigidify" it by attaching at each $a \in VA$ a rooted tree (T_a, a) with no non-trivial automorphism, no two of which are isomorphic. Example: $A = $ 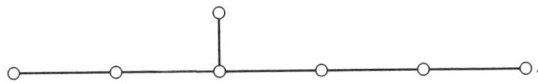 extended to 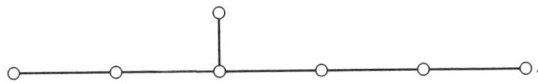. We can eliminate such rather trivial examples by ruling out terminal vertex (of degree 1). A finite tree A is π-rigid if only if $\mathrm{Aut}(A) = \{1\}$. For example:

Proposition 7.3. *Let A be a connected graph with no vertex of degree 1 and no non-trivial automorphism. If* $|VA| \leq 6$ *then A is isomorphic to B or C below.*

$$B = \quad = \quad = \quad ,$$

$$C = \quad = \quad = \quad .$$

Proof. Let A be as in the proposition. Clearly loops and multiple edges give rise to non-trivial automorphisms. Hence A must be simplicial. The proposition can be checked (with due patience) by constructing all connected simplicial graphs with n vertices, $2 \leq n \leq 5$, noting that they all admit non-trivial automorphisms, and then inspecting all ways of attaching a sixth vertex of degree d, $2 \leq d \leq 4$ (since a degree 5 vertex would give a cone, which inherits the symmetries of the base).

7.4 Connected simplicial graphs with $n \leq 5$ vertices

One can build them inductively, adding one new vertex at a time. For $n \geq 2$ they all have non-trivial symmetries.

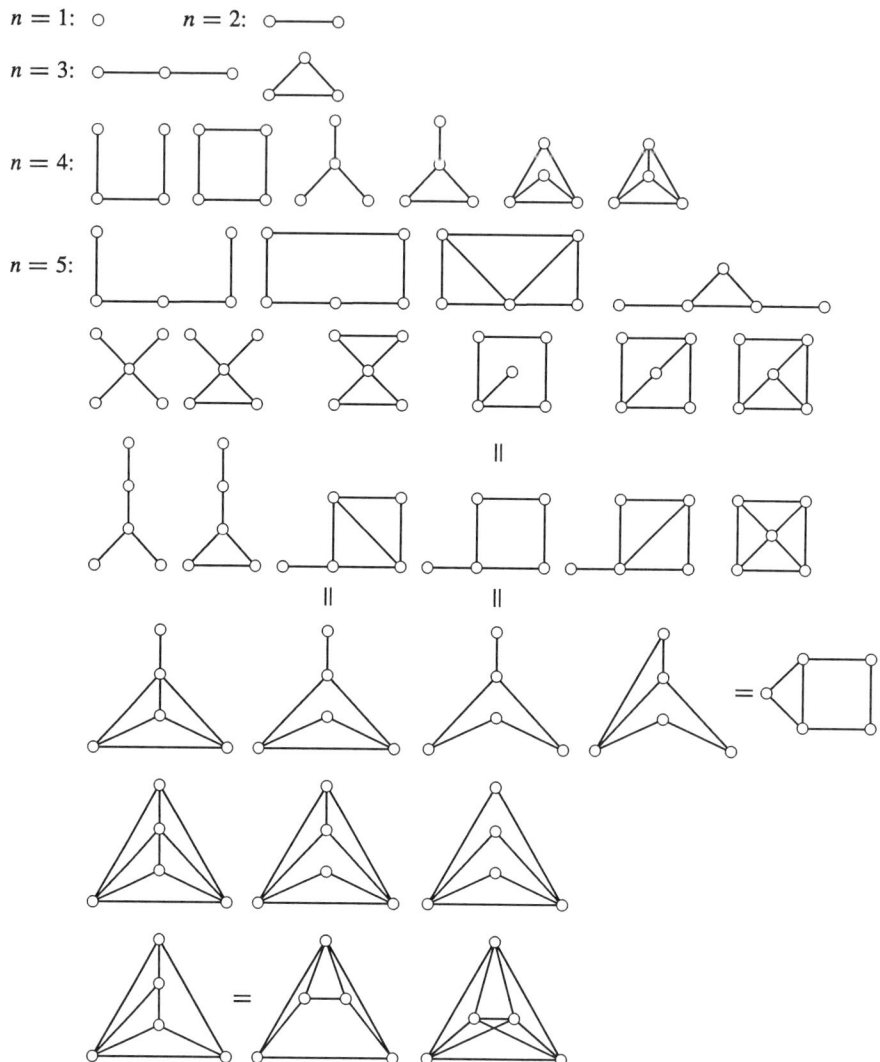

Proposition 7.5. *The graphs B and C of (7.3) are π-rigid, and hence they are the smallest (vertex size) π-rigid graphs without vertices of degree < 2.*

First consider

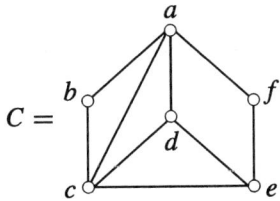

$$C =$$

Vertex: a b c d e f
Degree: 4   2   4   3   3   2
R_1-classes: $\{a, c\}$, $\{b, f\}$, $\{d, e\}$
$R_2 = Id$. Hence C is π-rigid.

Next consider $B = C - $ (the edge (c, e)).
Vertex: a b c d e f
Degree: 4   2   3   3   2   2
R_1-classes: $\{a\}$, $\{b, e, f\}$, $\{c, d\}$
R_2-classes: $\{a\}$, $\{b\}$, $\{e\}$, $\{f\}$, $\{c, d\}$
$R_3 = Id$. Hence B is π-rigid.

Thus, to get π-rigid examples without vertices of degree ≤ 2 we must go to at least 7 vertices.

Example 7.6. With seven vertices, none of degree ≤ 2.

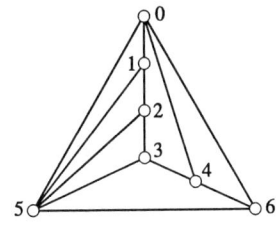

Vertex: 0   1   2   3   4   5   6
Degree: 4   3   3   3   3   5   3
R_1-classes: $\{0\}$, $\{1, 2, 3, 4, 6\}$, $\{5\}$
R_2-classes: $\{0\}$, $\{1, 6\}$, $\{2, 3\}$, $\{4\}$, $\{5\}$
R_3-classes: $\{0\}$, $\{1\}$, $\{2\}$, $\{3\}$, $\{4\}$, $\{5\}$, $\{6\}$.

Example 7.7. The initially presented example ([BK], (4.12)2) of a π-rigid graph is:

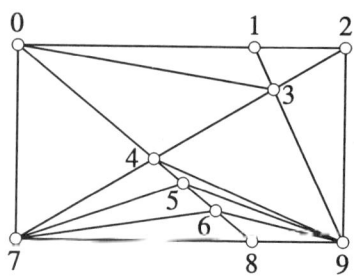

Vertex:	0	1	2	3	4	5	6	7	8	9
Degree:	4	3	3	5	5	4	4	5	3	6

R_1-classes: $\{0, 5, 6\}$, $\{1, 2, 8\}$, $\{3, 4, 7\}$, $\{9\}$
R_2-classes: $\{0\}$, $\{5\}$, $\{6\}$, $\{1\}$, $\{2\}$, $\{8\}$, $\{3\}$, $\{4\}$, $\{7\}$, $\{9\}$

References

[B] H. Bass, Covering theory for graphs of groups, *J. Pure Appl. Algebra*, **89** (1993), 3–47.

[BK] H. Bass and R. Kulkarni, Uniform tree lattices, *J. Amer. Math. Soc.*, **3** (1990), 843–902.

[BL] H. Bass and A. Lubotzky, *Tree Lattices*, Progress in Mathematics 176, Birkhäuser, Boston, 2000.

[L] F. T. Leighton, Finite common coverings of graphs, *J. Combin. Theory Ser. B*, **33** (1982), 231–238.

[S] J.-P. Serre, *Trees*, Springer-Verlag, Berlin, 1980.

[T] J. Tits, Sur le groupe des automorphismes d'un arbre, in A. Haefliger and R. Narasimhan, eds., *Essays in Topology and Related Topics*, Memories dedié à Georges de Rham, Springer-Verlag, Berlin, 1970, 188–211.

Appendix [PN]

The PNeumann Groups

Hyman Bass and Alexander Lubotzky

1 The construction and properties

1.0 The pair (S, V)

Following is a (slight variant on a) construction kindly supplied to us, on request, by Peter Neumann.

We start with a pair (S, V) consisting of

(1) $\qquad S =$ a simple inn-exact group (cf. (6.1)) of order m, and

(2) $\qquad V =$ a transitive rigid S-set of size $r > 1$.

Recall that "inn-exact" means that $ad : S \longrightarrow \mathrm{Aut}(S)$ is an isomorphism, i.e., $Z(S) = \{1\} = \mathrm{Out}(S)$. For V, transitive means $V \cong S/H$ for some $H \leq S$, and "rigid" means that $\mathrm{Aut}_S(V)$ ($\cong N_S(H)/H$) $= \{1\}$. This happens for example when V is primitive, i.e., H is maximal (and not normal, S being simple) in S. Possibilities for S include $M_{11}, M_{23}, M_{24}, J, E_8(p), F_4(p)$ $(p > 2), G_2(p)$ $(p > 2), PSp(2n, p)$ $(p > 2$ or $n > 2), \ldots$, where p is a prime.

The aim of this construction can be found in the summary at the end of this appendix.

1.1 Notation

If H is a group and W is a set, the product H^W denotes the group of all functions from W to H, and $H^{(W)}$ denotes the corresponding weak direct product, consisting of the functions with finite support; it is the subgroup generated by the factors $H_w (w \in W)$. For $1 \leq n \leq \infty$ we put $[n] = \{i \in \mathbf{Z} | 1 \leq i \leq n\}$, and $_0[n] = \{0\} \cup [n]$

1.2   The Construction

First form the product

(1) $$S^{o[\infty]} = S_0 \times S_1 \times S_2 \times S_3 \times \cdots$$

consisting of infinite sequences $s = (s_0, s_1, s_2, \ldots)$ of elements of S, S_n denoting the nth factor (all coordinates but the nth equal 1). Next for $n \geq 1$, we define

(2)   An $S^{o[\infty]}$ set $V_n = V \times V$, with action given by

$$s(v', v) = (s_{n-1}v', s_n v)$$

for $s = (s_0, s_1, s_2, \ldots) \in S^{o[\infty]}$. The S-rigidity of V implies the $(S_{n-1} \times S_n)$-rigidity of V_n.

(3)   For $J \subset [\infty]$ put $U_J = \coprod_{j \in J} V_j$ (disjoint union), an $S^{o[\infty]}$-set.

Then we can form the wreath product

(4) $$\Pi_{[\infty]} = S^{o[\infty]} \ltimes S^{U_{[\infty]}}.$$

For $1 \leq n < \infty$ this contains the finite group

(5)
$$\begin{aligned}
\Pi_{[n]} &= S^{o[n]} \ltimes S^{U_{[n]}} \\
&= (S_0 \times S_1 \times \cdots \times S_n) \ltimes (S^{(V_1 \sqcup \cdots \sqcup V_n)}).
\end{aligned}$$

Clearly $\Pi_{[n]} < \Pi_{[n+1]}$. Further define

(6)
$$\Delta_n : S \longrightarrow \Pi_{[\infty]}$$
$$\Delta_n(s) = (\Delta_n^0(s), \Delta_n^1(s)), \quad \Delta_n^0(s) \in S_0^{[\infty]}, \quad \Delta_n^1(s) \in S^{U_{[\infty]}},$$

where $\Delta_n^0(s)$ has ith coordinate 1 for $i \leq n$, and s for $i > n$; and the S^{V_i} coordinate of $\Delta_n^1(s)$ is 1 for $i \leq n$, and the constant function with value s for $i > n$.

Since $\Delta_n^0(s)$ and $\Delta_n^1(t)$, $(s, t \in S)$ (are easily seen to) commute, Δ_n is a (clearly injective) homomorphism. Figuratively,

$$\Delta_n(S) \text{ is the subgroup “diagonal after the } n\text{th factor.”}$$

We record some readily verified properties

(7)
$$\Delta_n : S \longrightarrow \Pi_{[\infty]} \text{ is an injective homomorphism.}$$
$$\Delta_n(S) \text{ centralizes } \Pi_{[n]}.$$
$$\Delta_n(S) \cap \Pi_{[m]} = \{1\} \quad \forall m < \infty.$$

We now put

(8) $$\Gamma_n := \langle \Delta_n(S), \Pi_{[n]} \rangle = \Delta_n(S) \times \Pi_{[n]}.$$

Clearly $\Delta_n(S) < \Gamma_{n+1}$, and so

(9) $$\Gamma_n < \Gamma_{n+1}.$$

Moreover,

(10) $$\begin{aligned} [\Gamma_{n+1} : \Gamma_n] &= [\Pi_{[n+1]} : \Pi_{[n]}] \\ &= |S_{n+1} \ltimes S^{V_{n+1}}| \\ &= |S|^{1+|V|^2} \\ &= q := m^{r^2+1}, \end{aligned}$$

a number independent of n. On the other hand

(11) $$\begin{aligned} |\Gamma_1| &= |\Delta_1(S) \times ((S_0 \times S_1) \ltimes S^{V_1})| \\ &= m^{r^2+3} = m^2 q. \end{aligned}$$

Finally, we put

(12) $$\Gamma_\infty = \bigcup_n \Gamma_n, \quad \text{and}$$

(13) $$\begin{aligned} \Pi_{(\infty)} &= \bigcup_n \Pi_{[n]} \\ &= S^{(\infty)} \ltimes S^{(U_{[\infty]})}, \quad \text{where } S^{(\infty)} := S^{(0_{[\infty]})} \end{aligned}$$

and we use the product notational conventions given above in (1.1).

We shall call the group Γ_∞ with its filtration $\Gamma_1 < \Gamma_2 < \Gamma_3 < \cdots$ the *PNeumann group(s) associated with* (S, V).

We have an exact sequence

(14) $$1 \longrightarrow \Pi_{(\infty)} \longrightarrow \Gamma_\infty \xrightarrow{\phi} S \longrightarrow 1$$

defined as follows: Let $p : \Pi_{[\infty]} \longrightarrow S^{0_{[\infty]}}$ denote the projection in (4). For $g \in \Gamma_\infty$, $p(g) = (s_0, s_1, s_2, \dots)$ is an eventually constant sequence: For some $s \in S$, $s_n = s$, for all $n \gg 0$. We put $\phi(g) = s$. Then the exact sequence (14) is readily checked, as is the fact that

(15) $$\Delta_n : S \longrightarrow \Gamma_\infty \text{ splits (14) for all } n \geq 1.$$

1.3   Normal subgroups

Let $L \subset {}_0[\infty]$ and $J \subset [\infty]$. We put

(1) $$\Pi_{(L,J)} = S^{(L)} \rtimes S^{(U_J)},$$

where $U_J = \coprod_{j \in J} V_j$ ((1.2)(3)), and write, for $1 \le n < \infty$,

(2) $L_{[n]} = L \cap [n]$, ${}_0J = \{0\} \cup J$,   and $(J-1)_n = (J-1) \cup \{n\}$.

Fix n.

(3)   For $L \subset {}_0[n]$ and $J \subset [n]$ put
$$J_L = L_{[n]} \cup (L+1)_{[n]} = (L \cup (L+1))_{[n]}, \quad \text{and}$$
$$L_J = {}_0J \cap (J-1)_n.$$

We claim that:

(4) $\Pi_{(L,J)} \lhd \Pi_{[n]}$ iff $J_L \subset J$ iff $L \subset L_J$. Moreover, every normal subgroup of $\Pi_{[n]}$ is of this form.

First note that

$$L \subset L_J \iff [L \subset {}_0J \quad \text{and} \quad L \subset (J-1)_n]$$
$$\iff [L_{[n]} \subset J \quad \text{and} \quad (L+1)_{[n]} \subset J]$$
$$\iff J_L \subset J.$$

1.4   Proof of (1.3)(4)

First observe that if $N \lhd S \times H$, H being any group, then $N = (N \cap S) \times (N \cap H)$. In fact the commutator $(N, S) \le N \cap S$ is either trivial, so $N \le H$, or equals S. Consequently,

(1)   Normal subgroups M of $\Gamma_n = \Delta_n(S) \times \Pi_{[n]}$ are of the form $M = \Delta_n(S) \times N$ or $M = N$, where $N \lhd \Pi_{[n]}$.

Further, by the same observation, for $N \lhd \Pi_{[n]}$, $N \cap S^{U_{[n]}} = S^W$ for some $W \subset U_{[n]}$. Moreover, W must be $S^{0[n]}$ invariant, hence $W = U_J$ for some $J \subset [n]$. Similarly, if

$$1 \longrightarrow S^{U_{[n]}} \longrightarrow \Pi_{[n]} \overset{p}{\longrightarrow} S^{0[n]} \longrightarrow 1$$

is the projection exact sequence, then we must have $p(N) = S^L$ for some $L \subset {}_0[n]$, $L = $ the "support" of $p(N)$. The action of $S^{0[n]}$ on $S^{U_{[n]}}$ is by permutation of factors, hence disjoint from inner automorphisms. It follows that, for $1 \le i \le n$,

$$(N, S^{V_i}) \le N \cap S^{V_i} \le S^{U_J}$$

is non-trivial iff either $i \in J$ or $(S^L, S^{V_i}) \neq \{1\}$. The latter happens iff $i \in L$ or $i - 1 \in L$ (see (1.2)(2)), i.e., $i \in J_L = L_{[n]} \cup (L+1)_{[n]}$. In this case we must have $i \in J$ also, whence $J_L \subset J$. The argument above shows that

(2) $\qquad\qquad\qquad (S^L, S^{U_{J_1}}) = \{1\} \quad \text{iff} \quad J_L \cap J_1 = \emptyset.$

Let $g \in N$ and write $g = p(g)g'$, with $p(g) \in S^L$ and $g' \in S^{U_{[n]}}$. Modifying g by an element of $S^{U_J} \leq N$, we can assume that $g' \in S^{U_{J'}}$, $J' = [n] - J$. Then the condition $J_L \subset J$ (verified above) implies (by (2)) that S^L centralizes $S^{U_{J'}}$. Taking commutators, we have

$$(g, S^{U_{J'}}) = (g', S^{U_{J'}}) \subset N \cap S^{U_{J'}} = \{1\}.$$

It follows that $g' = 1$, i.e., $g = p(g)$. This shows that $N = \Pi_{(L,J)}$. Similar calculations show that $\Pi_{(L,J)}$ is normal in $\Pi_{[n]}$ whenever $J_L \subset J$, whence (1.3)(4).

1.5   Consequences of (1.3)(4)

The following consequence is easily seen directly.

(1) $\qquad\qquad\qquad\qquad\qquad Z(\Gamma_n) = \{1\}.$

Moreover,

(2)   The minimal normal subgroups of Γ_n are

$$\Delta_n(S), S^{V_1}, \ldots, S^{V_n},$$

of orders $m, m^{r^2}, \ldots, m^{r^2}$. Among the normal subgroups of $\Pi_{[n]}$ not contained in $\mathrm{socle}(\Pi_{[n]}) = S^{U_{[n]}}$, the minimal ones are

$$S_0 \times S^{V_1}, S_1 \times S^{V_1 \sqcup V_2}, \ldots, S_{n-1} \times S^{V_{n-1} \sqcup V_n}, S_n \times S^{V_n}.$$

We next calculate centralizers.

(3)   Normal subgroups $\Pi_{(L,J)}$ and $\Pi_{(L_1,J_1)}$ of $\Pi_{[n]}$ centralize each other iff $J \cap J_1 = \emptyset$. Hence

$$Z_{\Pi_{[n]}}(\Pi_{(L,J)}) = \Pi_{(L_{J'},J')} \quad \text{(which depends only on } J),$$

where $J' = [n] - J$, and the double centralizer of $\Pi_{(L,J)}$ is therefore $\Pi_{(L_J,J)}$.

Clearly the condition $J \cap J_1 = \emptyset$ is necessary for centralization. Since $J_L \subset J$ we have then $J_L \cap J_1 = \emptyset$ also, so S^L centralizes $S^{U_{J_1}}$, by (2). Similarly S^{L_1} centralizes

S^{U_J}. It remains to see that S^L centralizes S^{L_1}, i.e., that $L \cap L_1 = \emptyset$. This follows since

$$L \cap L_1 \subset L_J \cap L_{J_1}$$
$$= ({}_0J \cap {}_0J_1) \cap ((J-1)_n \cap (J_1-1)_n)$$
$$= \{0\} \cap \{n\} = \emptyset.$$

Note now that, for $1 \le j \le n$, $L_{\{j\}} = {}_0\{j\} \cap \{j-1\}_n = \{0, j\} \cap \{j-1, n\}$, so

(4)

$$L_{\{j\}} = \begin{cases} \{0, 1\}, & 1 = j = n, \\ \{0\}, & 1 = j < n, \\ \{n\}, & 1 < j = n, \\ \emptyset, & 1 < j < n, \text{ and so} \end{cases}$$

$$Z_{\Pi_{[n]}}(Z_{\Pi_{[n]}}(S^{V_j})) = \begin{cases} (S_0 \times S_1) \bowtie S^{V_1}, & 1 = j = n, \\ S_0 \bowtie S^{V_1}, & 1 = j < n, \\ S_n \bowtie S^{V_n}, & 1 < j = n, \\ S^{V_j}, & 1 < j < n. \end{cases}$$

1.6 Automorphism groups

We next construct an (outer) involution

(1) $$\tau = \tau_n : \Pi_{[n]} \longrightarrow \Pi_{[n]}, \quad \tau^2 = Id,$$

as follows. On $S^{0[n]}$,

$$\tau((s_0, s_1, \ldots, s_{n-1}, s_n)) = (s_n, s_{n-1}, \ldots, s_1, s_0).$$

On $g \in S^{U_{[n]}}$, $g : U_{[n]} \longrightarrow S$, $\tau(g) = g \circ \tau'$, where $\tau' : U_{[n]} \longrightarrow U_{[n]}$ is defined by

$$\tau' : V_j \longrightarrow V_{n+1-j} \quad (1 \le j \le n),$$
$$\tau(v', v) = (v, v').$$

It is straightforward to verify that τ defines a homomorphism, and $\tau^2 = Id$. We now determine $\text{Aut}(\Gamma_n)$.

(2)

$$\text{Aut}(\Gamma_n) = \text{Aut}(\Delta_n(S)) \times \text{Aut}(\Pi_{[n]}) \text{ and } \Delta_n(S) \cong S \text{ is inn-exact.}$$
$$\text{Aut}(\Pi_{[n]}) = ad(\Pi_{[n]}) \bowtie \langle \tau_n \rangle \text{ with } \tau_n \text{ as in (1)}.$$

1.7 Proof of (1.6)(2)

$\Delta_n(S)$, the smallest normal subgroup of Γ_n (see (1.5)(2)), is characteristic, as is it's centralizer $\Pi_{[n]}$, whence the first assertions.

Let $\alpha \in \text{Aut}(\Pi_{[n]})$. The minimal normal subgroups (of $\Pi_{[n]}$) with the largest double centralizers are S^{V_1} and S^{V_n}, with double centralizers $M_0 = S_0 \times S^{V_1}$ and

$M_n = S_n \times S^{V_n}$ (see (1.5)(2) and (4)). Hence α either preserves or exchanges M_0 and M_n. Since τ exchanges them, we can modify α by τ, if necessary, and assume that α preserves them. It thus suffices to show that if $\alpha(M_0) = M_0$ then α is inner. Apart from M_0 and M_n, the normal subgroups minimal among those not contained in $S^{U_{[n]}}$ are $M_j = S_j \ltimes S^{V_j \sqcup V_{j+1}} (1 \le j < n)$ (see (1.5)(2)). The only one of these not centralizing M_0 is M_1, hence $\alpha(M_1) = M_1$. Among $M_2, \ldots, M_n$, the only one not centralizing M_1 is M_2, hence $\alpha(M_2) = M_2$. In this fashion we see that $\alpha(M_j) = M_j$ for all $j = 0, \ldots, n$. Since α leaves the socle, $S^{U_{[n]}}$, invariant, it follows that $\alpha(S^{V_j}) = S^{V_j} (1 \le j \le n)$, and α induces an automorphism of $S^{0[n]} = \Pi_{[n]}/S^{U_{[n]}}$ leaving each S_j invariant. Since S is inn-exact, we can modify α by an inner automorphism of $S^{0[n]}$ and assume that α induces the identity on $S^{0[n]}$. This done, the automorphism induced by α on S^{V_j} will be $S^{0[n]}$ equivariant. In particular the permutation of factors it induces will be an $S^{0[n]}$ automorphism of V_j. Since V_j is $S^{0[n]}$-rigid, by (1.2)(2), i.e., $\mathrm{Aut}_{S^{0[n]}}(V_j) = \{1\}$, α leaves invariant each factor ($\cong S$) of S^{V_j}. Since S is inn-exact, we can now further modify α by $ad(g)$ for some $g \in S^{U_{[n]}}$ and make α the indentity on $S^{U_{[n]}}$, as well as on $S^{0[n]} = \Pi_{[n]}/S^{U_{[n]}}$. In this case α has the form $\alpha(x) = \delta(x)x$, where $\delta : \Pi_{[n]} \longrightarrow Z(S^{U_{[n]}}) = \{1\}$, whence $\alpha = Id$, and (1.6)(2) is proved.

1.8 Relative automorphism groups

We next show that outer automorphisms of Γ_n do not prolong to Γ_{n+1}.

(1)
$$\text{(a) } N_{\Gamma_{n+1}}(\Gamma_n) = \Gamma_n,$$
$$\text{(b) } ad_{\Gamma_{n+1}} : \Gamma_n \xrightarrow{\approx} \mathrm{Aut}(\Gamma_{n+1}; \Gamma_n) \text{ is an isomorphism.}$$

To prove (a) first note that

$$N := \Pi_{(0[n-1],[n])} = S^{0[n-1]} \ltimes S^{U_{[n]}} < \Pi_{[n]}$$

is, by (1.3)(4), a normal subgroup of Γ_n. It therefore suffices to show that

$$\Gamma_n/N = \Delta_n(S) \times S_n \text{ is self-normalizing in}$$
$$\Gamma_{n+1}/N = \Delta_{n+1}(S) \times ((S_n \times S_{n+1}) \ltimes S^{V_{n+1}}),$$

Here $\Delta_n(S)$ is embedded in Γ_{n+1} as the diagonal subgroup of $\Delta_{n+1}(S) \times (S_{n+1} \ltimes S^{V_{n+1}})$, i.e., for $s \in S$, $\Delta_n(s)$ in Γ_{n+1}/N corresponds to

$$(\Delta_{n+1}(s), s, f) \in \Delta_{n+1}(S) \times S_{n+1} \times S^{V_{n+1}},$$

with $f : V_{n+1} \longrightarrow S$ the constant function of value s. Thus $Z_{\Gamma_{n+1}/N}(\Delta_n(S)) = S_n$, so $N_{\Gamma_{n+1}/N}(\Delta_n(S)) = \Delta_n(S) \times S_n(= \Gamma_n/N)$, since S is inn-exact. Inner automorphisms of Γ_{n+1}/N keep S_n inside the normal subgroup $(S_n \times S_{n+1}) \ltimes S^{V_{n+1}}$, and so cannot exchange S_n and $\Delta_n(S)$. Hence $N_{\Gamma_{n+1}/N}(\Gamma_n/N) \leq N_{\Gamma_{n+1}/N}(\Delta_n(S)) = \Gamma_n/N$, as observed above.

To prove (b), consider an $\alpha \in \mathrm{Aut}(\Gamma_{n+1}; \Gamma_n)$. If we show that α is inner, $\alpha = ad(g)$, then $g \in N_{\Gamma_{n+1}}(\Gamma_n) = \Gamma_n$, by (a). By (1.6)(2), α leaves $\Delta_{n+1}(S)$ and $\Pi_{[n+1]}$ invariant, and is inner on $\Delta_{n+1}(S)$. On $\Pi_{[n+1]}$, if α is not inner, then it follows from (1.6)(2) that $\alpha(S^{V_1}) = S^{V_{n+1}}$. This contradicts the assumption that $\alpha(\Gamma_n) = \Gamma_n$, whence (b).

From (1) we easily deduce that

(2)
$$\text{(a)} \quad \bigcap_{n \geq 1} N_{\Gamma_\infty}(\Gamma_n) = \Gamma_1 \quad \text{and}$$

$$\text{(b)} \quad ad_{\Gamma_\infty} : \Gamma_1 \xrightarrow{\cong} \mathrm{Aut}^{\mathrm{filt}}(\Gamma_\infty) \text{ is an isomorphism.}$$

In fact, let $\alpha \in \mathrm{Aut}^{\mathrm{filt}}(\Gamma_\infty)$. Then $\alpha|\Gamma_{n+1} \in \mathrm{Aut}^{\mathrm{filt}}(\Gamma_{n+1}; \Gamma_n)$, so, by (1), there is a unique $g_n \in \Gamma_n$ such that $\alpha = ad(g_n)$ on Γ_{n+1}. Then $ad(g_n)$ and $ad(g_{n+1})$ agree on Γ_{n+1}, so $g_n^{-1} g_{n+1} \in Z(\Gamma_{n+1}) = \{1\}$ ((1.5)(1)). Thus $g_n = g_1 \in \Gamma_1$ for all $n \geq 1$, whence (2).

1.9 Remark

Peter Neumann's original construction gave slightly smaller groups Γ'_n, where S_0 was not present, and, in place of $V_1 = V \times V$, one had $V'_1 = V$, acted upon by S_1. With these changes essentially everything above goes through, with the added bonus that Γ'_n lacks the bilateral symmetry of Γ_n, expressed by τ_n, so that Γ' is even inn-exact.

On the other hand, $\Gamma'_1 = \Delta_1(S) \times (S_1 \ltimes S^{V'_1})$ has a normal subgroup $(S_1 \ltimes S^{V'_1}) \lhd \Gamma'_\infty$, and this is a problem for our applications. Moreover, Γ'_1, of order m^{r+2}, does not contain a subgroup of index $q = m^{r^2+1}$, with which we could try to repair the situation.

In contrast, in $\Gamma_1 = \Delta_1(S) \times ((S_0 \times S_1) \times S^{V_1})$, we have the subgroup

(1) $\Gamma_0 := \Delta_1(S) \times S_0$, of index $q = m^{r^2+1}$ in Γ_1, and Γ_0 contains no non-trivial normal subgroup of Γ_n for any $n \geq 1$.

Moreover the proof of (1.8)(1) applies verbation to show that

(2)
$$\text{(a)} \quad N_{\Gamma_1}(\Gamma_0) = \Gamma_0 \quad \text{and}$$

$$\text{(b)} \quad ad_{\Gamma_1} : \Gamma_0 \xrightarrow{\cong} \mathrm{Aut}(\Gamma_1; \Gamma_0) \text{ is an isomorphism.}$$

For (b) we need only observe that $\mathrm{Aut}(\Gamma_0) = ad(\Gamma_0) \rtimes \langle \tau_0 \rangle$, where τ_0 is the involution switching $\Delta_1(S)$ and S_0, and τ_0 does not extend to $\mathrm{Aut}(\Gamma_1)$. Now combining (1.8)(2) and (2), we obtain:

(3)   (a) $\bigcap_{n \geq 0} N_{\Gamma_\infty}(\Gamma_n) = \Gamma_0$, and

       (b) $ad_{\Gamma_\infty} : \Gamma_0 \xrightarrow{\cong} \text{Aut}^{\text{filt}^+}(\Gamma_\infty)$ is an isomorphism, where filt^+ refers to the filtration $(\Gamma_n)_{n \geq 0}$.

1.10   Normal subgroups of Γ_∞

For $L \subset {}_0[\infty]$ and $J \subset [\infty]$, we have

$$\Pi_{(L,J)} = S^{(L)} \ltimes S^{(U_J)}$$

as in (1.3)(1). Letting $n \to \infty$ in (1.3)(4), we conclude that

(1) $\Pi_{(L,J)} \lhd \Pi_{(\infty)}$ iff $L \cup (L+1) \subset {}_0 J := \{0\} \cup J$, in which case $\Pi_{(L,J)} \lhd \Gamma_\infty$. Moreover, every normal subgroup of $\Pi_{(\infty)}$ is of this form.

In fact, $\Pi_{(L,J)}$ above is normal in the full product group, $\Pi_{[\infty]} = S^{0[\infty]} \ltimes S^{U_{[\infty]}}$. Further:

(2)   If $M \lhd \Gamma_\infty$ and $M \not\leq \Pi_{(\infty)}$ then

$$M \cap \Pi_{(\infty)} = \Pi_{(L,J)}$$

with $L \subset {}_0[\infty]$ and $J \subset [\infty]$ cofinite, and $L \cup (L+1) \subset {}_0 J$. For sufficiently large n,

$$M = \Delta_n(S) \cdot \Pi_{(L,J)}.$$

In particular, Γ_∞ / M is finite.

1.11   Proof of (1.10)(2)

For sufficiently large n, $M \cap \Gamma_n \not\leq \Pi_{[n]}$, and so, by (1.4)(1), $\Delta_n(S) \leq M$, and in fact $\Delta_m(S) \leq M$ for all $m \geq n$. For $s \in S$, writing

$$\Delta_n(s) \in \Gamma_m = \Delta_m(S) \times (S^{0[m]} \ltimes S^{U_{[m]}})$$

in coordinates, $\Delta_n(s) = (\Delta_m(s), p(s), u(s))$ we see that the support of $p(s)$ in $S^{0[m]}$ is $(n, m] = \{n+1, \ldots, m\}$, i.e., $p(s) \in S_{n+1} \times \cdots \times S_m$, and no smaller product contains $p(s)$. Since $M \cap \Gamma_m \not\leq \Pi_{[m]}$ we have $\Delta_m(S) \leq M$, so

$$(p(s), u(s)) \in M \cap \Pi_{[m]} =: \Pi_{(L_m, J_m)},$$

as in (1.3)(4). It follows then that $(n, m] \subset L_m$ for all $m \geq n$. Letting $m \to \infty$ we see that $(n, \infty) := \bigcup_{m \geq n}(n, m] \subset \bigcup_{m \geq n} L_m = L$, so L is cofinite. Since $L \cup (L+1) \subset J$, J is likewise cofinite. This completes the proof of (1.10)(2).

We conclude with a summary of some of the properties of the Γ_n established above.

2   The PNeumann group: Summary

2.1   Theorem

Let S be a simple inn-exact group of order m, and V a transitive rigid S-set of size $r > 1$. Put $q = m^{r^2+1}$. Then there is a chain

(1)
$$\{1\} < \Gamma_0 < \Gamma_1 < \Gamma_2 < \Gamma_3 < \cdots, \quad \Gamma_\infty = \bigcup_n \Gamma_n,$$

with the following properties, for all $n \geq 0$.

(2)
$$|\Gamma_0| = m^2 \quad and \quad [\Gamma_{n+1} : \Gamma_n] = q.$$

(3)
Γ_n *has all composition factors* $\cong S$.
Γ_∞ *is residually finite, with the same property.*

(4)
$$Z(\Gamma_n) = \{1\}, \quad and \quad \mathrm{Aut}(\Gamma_n) = ad(\Gamma_n) \rtimes \langle \tau_n \rangle \quad with \ \tau_n^2 = Id.$$

(5)
(a) $N_{\Gamma_{n+1}}(\Gamma_n) = \Gamma_n$, *and*

(b) $ad_{\Gamma_{n+1}} : \Gamma_n \xrightarrow{\cong} \mathrm{Aut}(\Gamma_{n+1}; \Gamma_n)$.

(6)
(a) $\bigcap_{n \geq 0} N_{\Gamma_\infty}(\Gamma_n) = \Gamma_0$, *and*

(b) $ad_{\Gamma_\infty} : \Gamma_0 \xrightarrow{\cong} \mathrm{Aut}^{\mathrm{filt}}(\Gamma_\infty)$, *where "filt" refers to the filtration* $(\Gamma_n)_{n \geq 0}$.

Further we have an explicit description of all normal subgroups Γ_n, $0 \leq n \leq \infty$. (See (1.3)(4), (1.4)(1), (1.10)(1) and (2).) In particular,

(7) Γ_0 *contains no non-trivial normal subgroup of* Γ_n *for any* n, $1 \leq n \leq \infty$.

References

[AB]   R. C. Alperrin and H. Bass, Length functions of group actions on Λ-trees, in S. Gersten and J. Stallings, eds., *Combinatorial Group Theory and Topology*, Annals of Mathematics Studies 111, Princeton University Press, Princeton, NJ, 1987, 265–378.

[B1]   H. Bass, Some remarks on group actions on trees, *Comm. Algebra*, **4** (1976), 1091–1126.

[B2]   H. Bass, Group actions on non-archimedean trees, in *Arboreal Group Theory*, MSRI Publications 9, Springer-Verlag, Berlin, 1991, 69–131.

[B3]   H. Bass, Covering theory for graphs of groups, *J. Pure Appl. Algebra*, **89** (1993), 3–47.

[B4]   H. Bass, The Ihara-Selberg zeta function of a tree lattice, *Internat. J. Math.*, **3** (1992), 717 797.

[BCR]   H. Bass, L. Carbone, and G. Rosenberg, The Existence Theorem for tree lattices, appendix to H. Bass and A. Lubotzky, *Tree Lattices*, Birkhäuser, Boston, 2000, 167–184 (this volume).

[BJ]   H. Bass and R. Jiang, Automorphism groups of tree actions and of graphs of groups, *J. Pure Appl. Alegebra*, **112** (1996), 109–155.

[BK]   H. Bass and R. Kulkarni, Uniform tree lattices, *J. Amer. Math. Soc.*, **3** (1990), 843–902.

[BL]   H. Bass and A. Lubotzky, Rigidity of group actions on locally finite trees, *Proc. London Math. Soc.*, **69** (1994), 541–575.

[BORT] H. Bass, M. V. Otero-Espinar, D. N. Rockmore, and C. Tresser, *Cyclic Renormalization and Automorphisms of Rooted Trees*, Lecture Notes in Mathematics 1621, Springer-Verlag, Berlin, New York, Heidelberg, 1995.

[Be]   H. Behr, Arithmetic groups over finite fields I: A complete characteriza-
          tion of finitely generated and finitely presented arithmetic subgroups of
          reductive algebraic groups, *J. Reine Angew. Math.*, **485** (1998), 79–118.

[BG]   N. Benakli and Y. Glasner, Automorphism groups of trees acting locally
          with affine permutations, *Geom. Dedicata*, to appear.

[BH]   A. Borel and G. Harder, Existence of discrete cocompact subgroups of
          reductive groups over local fields, *J. Reine Angew. Math.*, **298** (1978), 53–
          64.

[BM1]   M. Burger and S. Mozes, CAT(-1) spaces, divergence groups and their
          commensurators, *J. Amer. Math. Soc.*, **9** (1996), 57–93.

[BM2]   M. Burger and S. Mozes, Finitely presented simple groups and product of
          trees, *C. R. Acad. Sci. Paris Sér. I Math.*, **324** (1997), 747–752.

[BM3]   M. Burger and S. Mozes, Groups acting on trees: From local to global
          structure, preprint IHES/M/99115, 1999.

[BM4]   M. Burger and S. Mozes, Lattices in products of trees, preprint
          IHES/M/99137, 1999.

[BMZ]   M. Burger, S. Mozes, and R. J. Zimmer, Irreducible lattices in the auto-
          morphism group of a product of trees, superrigidity and arithmeticity, in
          preparation.

[C1]   L. Carbone, Non-uniform lattices on uniform trees, *Mem. Amer. Math.
          Soc.*, to appear, 2000.

[C2]   L. Carbone, Non uniform lattices on uniform trees, Dissertation, Columbia
          University, New York, 1997.

[C3]   L. Carbone, Non-minimal actions and the existence of non-uniform tree
          lattices, in preparation, 2000.

[C4]   L. Carbone, Constructing tree lattices, in K. P. Shum and E. Taft, eds.,
          Algebras and Combinatorics: An International Congress (Proceedings of
          ICAC'97, Hong Kong, 1997), Springer-Verlag, Berlin, New York, Heidel-
          berg, 1999, 63–97.

[C5]   L. Carbone, Deformations of lattices in rank 1 groups of p-adic type,
          preprint, 2000.

[CG1]   L. Carbone and H. Garland, Lattices in Kac Moody groups, *Math. Res.
          Lett.*, **6** (1999), 439–447.

[CG2]   L. Carbone and H. Garland, Existence of lattices in Kac Moody groups over finite fields, in preparation, 2000.

[CR1]   L. Carbone and G. Rosenberg, Lattices on non-uniform trees, in preparation, 2000.

[CR2]   L. Carbone and G. Rosenberg, Infinite towers of tree lattices, in preparation, 2000.

[CR3]   L. Carbone and G. Rosenberg, Infinite towers of non-uniform tree lattices, in preparation, 2000.

[FTN]   A. Figa-Talamanca and C. Nebbia, *Harmonic Analysis and Representation Theory for Groups Acting on Trees*, London Mathematical Society Lecture Notes Series 162, Cambridge University Press, Cambridge, 1991.

[G1]   Y. Glasner, Some remarks on the covolume of lattices acting on a product of trees, Master's thesis, Hebrew University, Jerusalem, 1997.

[G2]   Y. Glasner, A two-dimensional version of the Goldschmidt–Sims conjecture, preprint.

[Le]   I. Levich, unpublished notes, Hebrew University, Jerusalem, 1995.

[Lei]   F. T. Leighton, Finite common coverings of graphs, *J. Combin. Theory B*, **33** (1982), 231–238.

[Li]   Y. S. Liu, Density of the commensurability groups of uniform tree lattices, *J. Algebra*, **165** (1994), 246–359.

[Li2]   Y. S. Liu, A necessary condition for an elliptic element to belong to a uniform tree lattice, *Proc. Amer. Math. Soc.*, **120** (1994), 1035–1039.

[Li3]   Y. S. Liu, A counterexample to the deformation conjecture for uniform tree lattices, *Proc. Amer. Math. Soc.*, **123** (1995), 315–319.

[L1]   A. Lubotzky, Trees and discrete subgroups of Lie groups over local fields, *Bull. Amer. Math. Soc.*, **20** (1989), 27–31.

[L2]   A. Lubotzky, Lattices of minimal covolume in SL_2: A nonarchimedean analogue of Sigel's Theorem $\mu \geq \pi/21$, *J. Amer. Math. Soc.*, **3** (1990), 961–975.

[L3]   A. Lubotzky, Lattices in rank one Lie groups over local fields, *Geom. Funct. Anal.*, **4** (1991), 405–431.

[L4]   A. Lubotzky, Tree lattices and lattices in Lie groups, in A. J. Duncan, N. D. Gilbert and J. Howic, eds., *Combinatorial and Geometric Group Theory*, London Mathematical Society Lecture Notes Series 204, Cambridge University Press, Cambridge, 1995, 217–232.

[LM]   A. Lubotzky and S. Mozes, Asymptotic properties of unitary representations of tree automorphisms, in M. A. Picardello, ed., *Harmonic Analysis and Discrete Potential Theory*, Plenum Press, New York, 1992, 289–298.

[LMZ]   A. Lubotzky, S. Mozes, and R. J. Zimmer, Superrigidity for the commensurability group of tree lattices, *Comm. Math. Helv.*, **69** (1994), 523–548.

[LW]   A. Lubotzky and T. Weigel, Lattices of minimal covolume in SL_2 over nonarchimedean fields, *Proc. London Math. Soc.*, **78** (1999), 283–333.

[M1]   S. Mozes, On the congruence subgroup problem for tree lattices, in S. G. Dani, ed., *Proceedings of Workshop on Ergodic Theory and Lie Groups* (Tata Institute of Fundmental Research, Mumbai, India, 1996), Narosa Publishing House, New Delhi, 1999, 143–149.

[M2]   S. Mozes, Trees, lattices and commesurators, in T. Y. Lam and A. R. Magid, eds. *Algebra, K-Theory, Groups, and Education*, Contemporary Mathematics 243, AMS, Providence, 1999, 141–151.

[M3]   S. Mozes, Products of trees, lattices and simple groups, in *Proceedings of the ICM, Berlin*, Vol. II, 1998, 571–582.

[Mar]   G. A. Margulis, *Discrete Subgroups of Semisimple Lie Groups*, Springer-Verlag, Berlin, New York, Heidelberg, 1989.

[Rag1]   M. S. Raghunathan, *Discrete Subgroups of Lie Groups*, Springer-Verlag, Berlin, New York, Heidelberg, 1972.

[Rag2]   M. S. Raghunathan, Discrete subgroups of algebraic groups over local fields of positive characteristics, *Proc. Indian Acad. Sci. Math. Sci.*, **99** (1989), 127–146.

[Se]   J. P. Serre, *Trees*, Springer-Verlag, New York, 1980.

[T]   T. Tamagawa, On discrete subgroups of p-adic algebraic groups, in O. F. G. Schilling, ed., *Arithematical Algebraic Geometry*, Harper and Row, New York, 1965, 11–17.

[Ti1]   J. Tits, Sur le groupe des automorphismes d'un arbre, in A. Haefliger and R. Narasimhan, eds., *Essays on Topology and Related Topics: Memories Dediés à George de Rham*, Springer-Verlag, Berlin, 1970, 188–211.

[Ti2]   J. Tits, A theorem of Lie-Kolchin for trees, in H. Bass, P. Cassidy, and J. Kovacik, eds., *Contributions to Algebra: A Collection of Papers Dedicated to Ellis Kolchin*, Academic Press, New York, 1977, 377–388.

Index of Notation

Index